Core Geography

2nd edition

Hodder & Stoughton
MEMBER OF THE HODDER HEADLINE GROUP

910

ACKNOWLEDGEMENTS

The authors and publishers would like to thank the following for permission to reproduce materials in this book. Every effort has been made to trace and acknowledge all copyright holders, but if any have been overlooked the publishers will be pleased to make the necessary arrangements.

Department of the Environment, Figure 11.13; The Guardian, Figures 9.16, 11.5, 11.10; Reed Consumer Books and George Philip, *Philip's Geographical Digest 1994–95*, Figures 7.5, 9.2; The Times © and Susan Elliot, Figure 1.13.

The publishers would also like to thank the following for giving permission to reproduce copyright photographs in this book.

Dan Addelman, Figures 5.3, 8.15, 10.2; AGF, Figure 10.6; Gene Ahrens, Figure 2.10; J Allan Cash Ltd, Figures 2.14, 6.14b, 9.1a; Associated Press Ltd, Figures 1.12, 1.16, 3.16; Bruce Coleman, Figures 2.12, 5.4, 11.6; Colorific, Figure 5.15; Colorpix, Figure 8.15; Colorama, Figure 4.9; Sue Cunningham, Figure 9.1b; Douglas Dickens, Figure 7.22; Ben Edwards/Impact, Figure 8.4; Greg Evans, Figures 4.1c, 4.12; Eye Ubiquitous, Figures 10.13a,b; Fiat Cars, Figure 9.9; GSF Picture Library, Figure 1.18; Robert Harding, Figures 4.8, 4.11, 4.15, 5.10a,b,c, 5.11, 5.21, 5.23, 6.14, 7.3, 7.4, 7.7, 8.3, 8.5, 8.10, 10.2b, 10.7b, 11.14; Robert Harding/Adam Woolfit, Figure 6.12a; Robert Harding/Rob Cousins, Figure 11.12; Honda, Figure 7.20; Hulton Deutsch, Figure 5.16; Hutchison Library, Figures 2.14, 5.18, 5.22, 6.5, 9.1c; The Independent/David Nichelson-Lord, Figure 10.16; Life File, Figures 1.9 (Gina Green), 2.8a(Eric Wilkins),b(Emma Lee), 5.12 (Jeremy Hoare), 8.12 (Mo Khan); Tony Morrison, Figures 4.4, 4.5; National Motor Museum, Figure 9.13; Popperfoto, Figures 1.1b, 1.6, 1.11, 2.18, 3.16; Roger Scruton/Impact, Figure 3.14; Spectrum, Figure 11.1; Christine Starkey, Figure 1.17; Telegraph Colour Library, Figure 6.15.

All other photos belong to the authors.

British Library Cataloguing in Publication Data

ISBN 0 340 74694 7

First edition published 1995

Impression number 10 9 8 7 6 5 4 3 2 1
Year 2005 2004 2003 2002 2001 2000

Typeset by Wearset, Boldon, Tyne and Wear.
Printed in Italy for Hodder & Stoughton Educational, a division of Hodder Headline Plc, 338 Euston Road, London NW1 3BH by Printer Trento.

CONTENTS

unit

The Violent Earth

Key questions

- Where do earthquakes and volcanoes happen, and why?
- What are earthquakes and volcanoes like?
- How do earthquakes affect countries in different states of economic development?
- Can anything be done to make earthquakes and volcanoes less of a threat?

The structure of the earth

The earth is made up of three main layers: the **crust**; the **mantle**; and the **core** (Figure 1.1). The crust is solid rock and varies in thickness from 6 km under the oceans to 70 km under mountain chains like the Himalayas. The mantle is thick, molten rock. The core is made up of an outer layer of very hot, liquid rock and an inner layer of iron.

The crust is very much thinner than the other layers of the earth and it is broken up into a number of pieces called **plates**. Huge currents of molten rock rise from the mantle and move the plates about, very slowly. The continents are passengers on these plates and so they are moved around as well; for example, you can see on a world map that South America and Africa used to fit together like the pieces of a jigsaw puzzle.

Figure 1.1 The structure of the earth

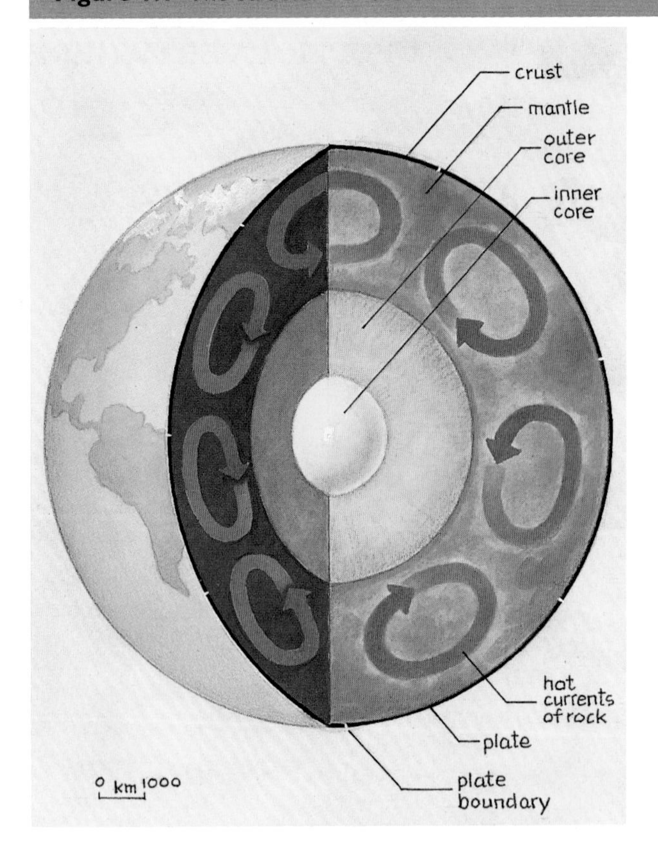

The distribution of earthquakes and volcanoes

Figure 1.2 shows the plates which make up the earth's crust and the places where earthquakes and volcanoes occur. It is easy to see that most earthquakes happen near to the **plate boundaries** where the plates are moving into, pulling away from or rubbing past each other. Stress builds up until the crust snaps apart, causing the ground to shake.

Most volcanoes also happen near to the plate boundaries. Where plates are moving together one piece of the crust is forced down into the mantle and melts; this molten material pushes its way back to the surface to form a volcano (see Figure 1.3). Where plates are moving apart, molten rock rises up to fill in the gap.

Figure 1.2 Plates, earthquakes and volcanoes

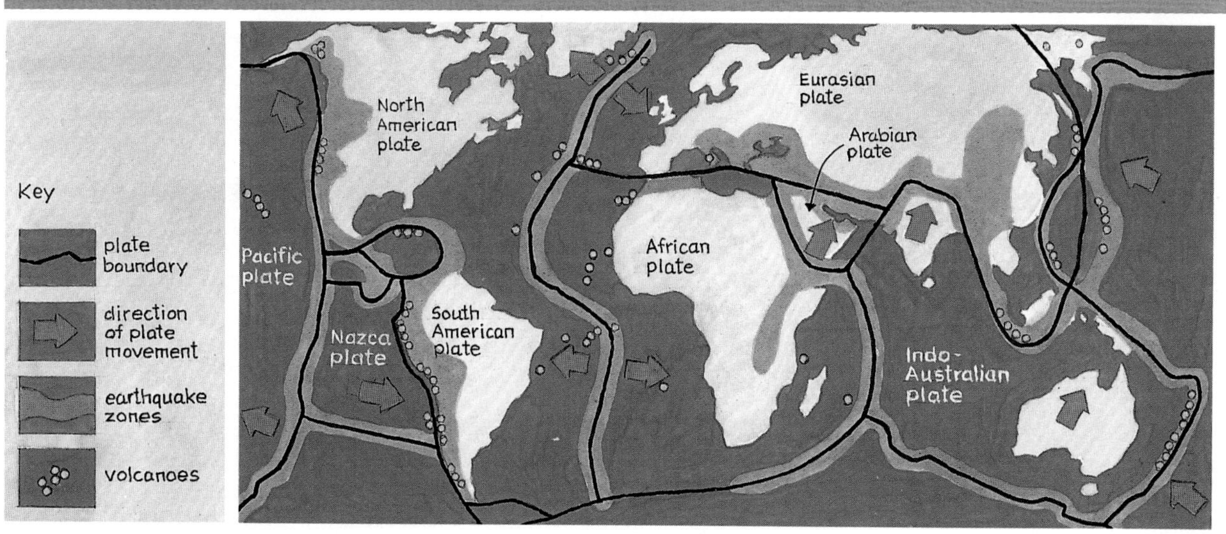

Figure 1.3 Plate boundaries and volcanoes

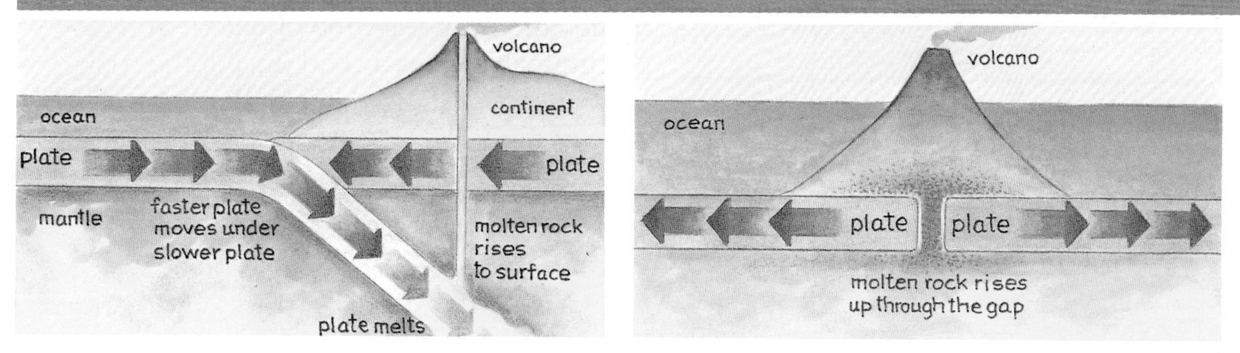

However, some volcanoes, like those which make up the Hawaiian Islands in the Pacific Ocean, happen a long way from the plate boundaries where there is a **hot spot** in the crust. The cause of these hot spots is still a mystery.

Word box

crust The layer of solid rock at the earth's surface

mantle The layer of molten rock beneath the crust

core The layers of liquid rock and solid iron at the earth's centre

plate A piece of the earth's crust

plate boundary The place where two plates meet

hot spot A place a long way from a plate boundary where volcanoes happen

QUESTION BOX

1. Which plate is the British Isles part of?
2. Look at Figure 1.2. Describe the pattern of earthquake and volcanoes.
3. Suggest why earthquakes and volcanoes are found at, or near, plate boundaries.

A Describe and explain the different types of plate movement.

B Why is the risk of earthquakes greater in Japan than in the British Isles?

Extension task: see worksheet 1A/B

The nature of earthquakes and volcanoes

Earthquakes

An earthquake is a shock wave passing through the ground. Small earthquakes can, for example, be caused by an underground cave collapsing but most earthquakes are caused when a piece of the earth's crust snaps as the plates move about. The place inside the crust where an earthquake starts is called the **focus** and the place on the surface directly above the focus is called the **epicentre** (Figure 1.4).

Figure 1.4 Earthquake: focus, epicentre and shock waves

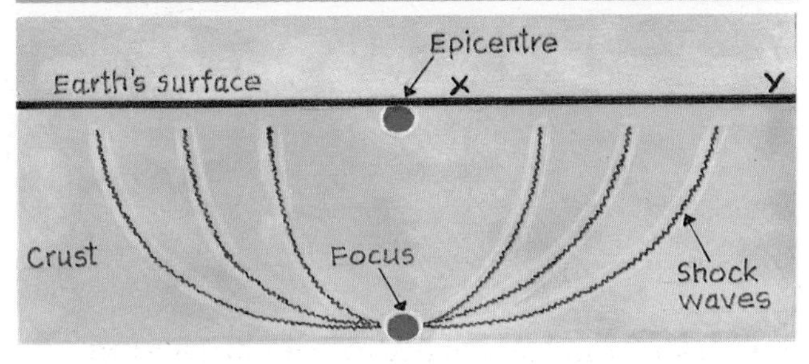

Two different scales are used to measure the strength of an earthquake. The Mercalli scale describes the damage caused by an earthquake while the Richter scale is a measure of the energy released by an earthquake when the rocks snap apart. Figure 1.5 shows how the two scales compare.

Figure 1.5 The Mercalli and Richter scales

Mercalli	Description	Richter
1		0
2	slight tremor	3.5
3		4.2
4	windows rattle	4.3
5		4.8
6	cupboards fall and trees sway	4.9
7		5.5
8	chimneys fall, roads crack	6.2
9		6.9
10	houses fall, huge cracks appear in ground	7.0
11		7.4
12	everything is destroyed	8.9

Figure 1.6 Tsunamis damage, Okushiri in Japan, July 1993

The damage caused by an earthquake depends on many things: for example, its strength; how close its epicentre is to where people live; how well the buildings have been made; and even the time of day (if an earthquake strikes a road during the rush hour it causes more damage than at a quieter time of day).

Other earthquake hazards are **tsunamis**. These are giant sea waves started by an earthquake in the crust under the ocean. They can travel at over 100 km per hour and be over 30 m high. For example, in July 1993 an earthquake measuring 7.8 on the Richter scale happened 80 km off Okushiri in the Sea of Japan. Tsunamis left 80 people dead and thousands homeless (Figures 1.6 and 1.7).

Figure 1.7 Location map: Okushiri, Japan

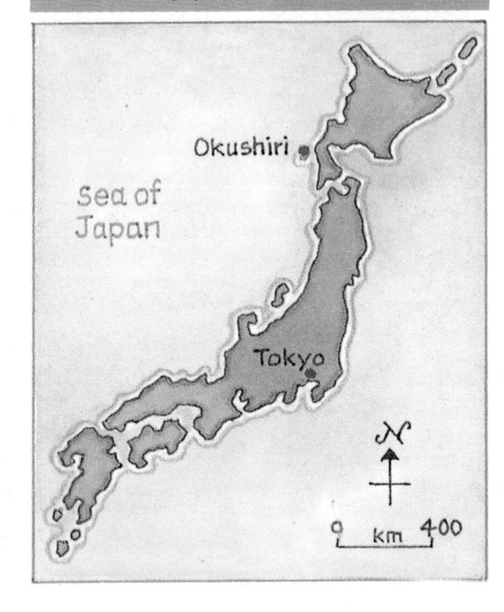

Volcanoes

A volcano is a crack in the earth's crust out of which comes lava, ash, steam and gas. The lava collects inside the crust in a magma chamber and when the pressure is great enough (imagine shaking up a can of fizzy drink) the volcano erupts. The pipe through which the lava comes is the **vent** and the hole at the top of the vent is the **crater**. The mound made by the lava is the **cone**.

Figure 1.8 Three types of volcano

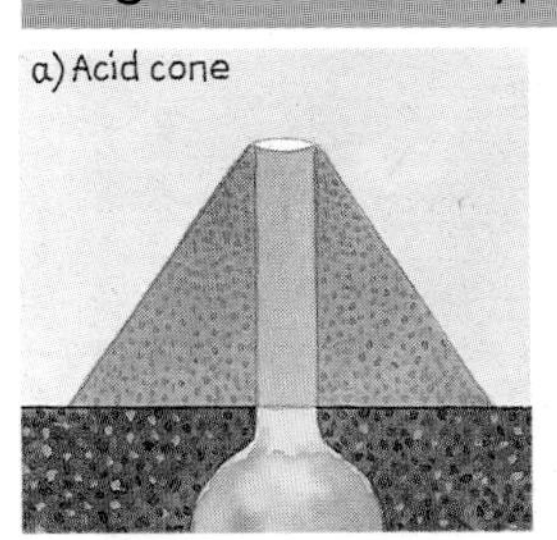

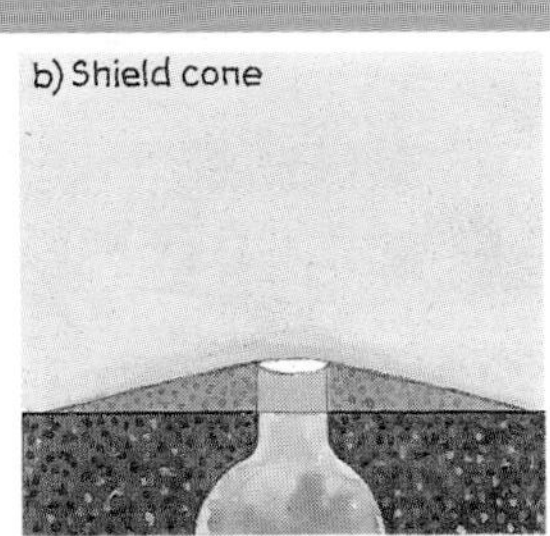

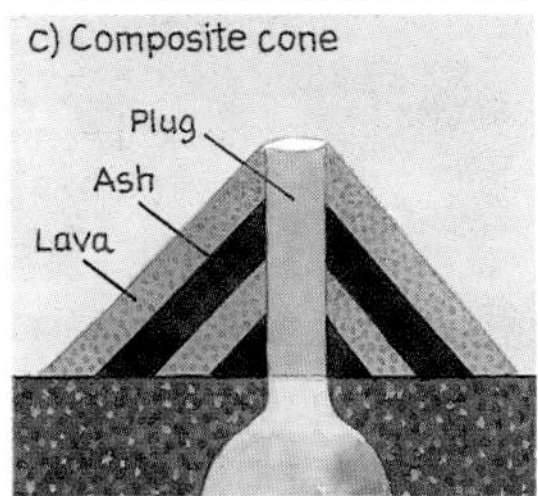

There are many different types of volcano, three of which are shown in Figure 1.8. Acid volcanoes have thick, slow moving lava and, as a result, they have steep slopes. Shield volcanoes have thin, runny lava and gentle slopes. Composite volcanoes have layers of lava and ash, one on top of the other. These form when a solid plug of lava is blasted out of the vent of a volcano, the fragments settling as ash; the lava can then flow until it cools down, solidifies and forms a layer on top of the ash and then another plug.

Volcanoes destroy roads, houses and farmland but usually few people are killed because most eruptions start slowly so there is time to get away. However, there are exceptions; for example, in 1902 Mont Pelée, on the island of Martinique in the West Indies, erupted with a huge explosion of fast moving gas, ash and lava which killed 30 000 people in a matter of minutes.

Figure 1.9 Hot spring at Kyushu, Japan

Volcanoes bring benefits as well as problems. For example, some types of lava make a very fertile soil; light falls of ash are a very good fertiliser; and they are a tourist attraction. Also, hot springs are found near many volcanoes and these can be used for hot baths or **geothermal power**; for example, in Japan there are several hundred hot spring resorts (Figure 1.9).

Word box

focus The place in the crust where an earthquake starts
epicentre The place on the surface above the focus of an earthquake
tsunami A sea wave caused by an earthquake
vent The pipe inside the volcano which the lava comes out through
crater The hole at the top of a volcano
cone The mound made by a volcano
geothermal power Energy which is generated (made) from hot rocks, or boiling water, or steam from inside the earth's crust

QUESTION BOX

1 Draw an example of a volcano and label onto it the following features: vent; crater; cone; magma chamber; and lava flow.

2 Look at Figure 1.4. Why would place X suffer more damage than place Y?

A Explain some of the reasons why people continue to live near volcanoes, despite the dangers.

B Why would an earthquake in Tokyo probably cause a greater loss of life if it struck at 8.00 am rather than 11.00 am?

Extension task: see worksheet 1C/D

CASE STUDY

The Los Angeles earthquake, 1994

On Monday 17 January 1994 at 4.40 am an earthquake measuring 6.6 on the Richter scale struck Los Angeles. Its epicentre was 32 km north west of the city centre. It lasted 45 seconds and it was followed by a number of **aftershocks**, the biggest of which measured 5.5 on the Richter scale.

Figure 1.11 Los Angeles earthquake damage in Slymar, 1994

Figure 1.10 Earthquake country, southern California

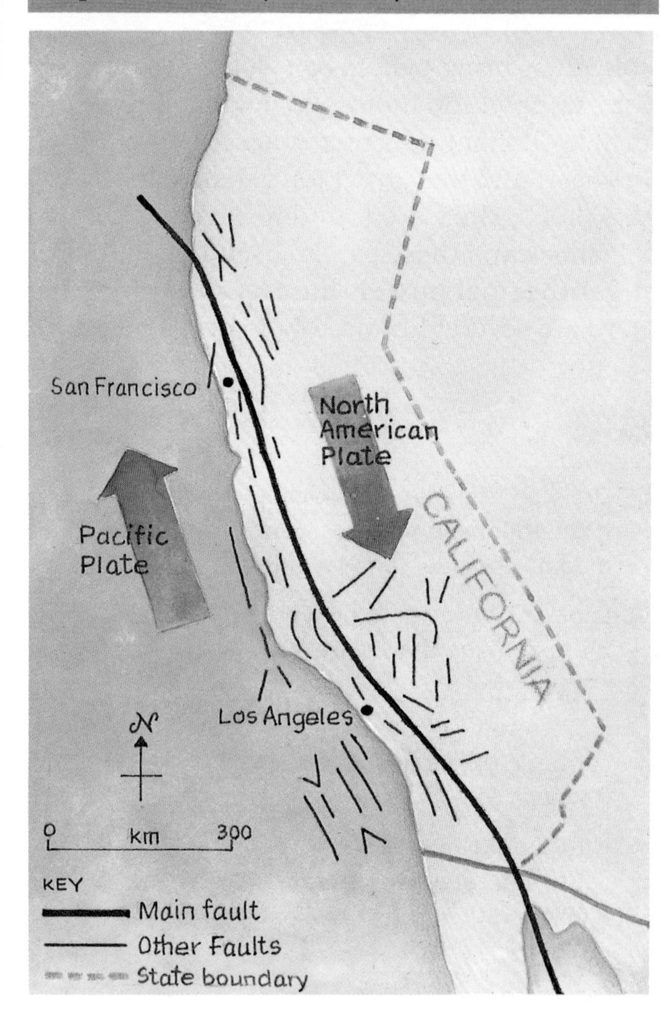

Los Angeles is in an earthquake zone because it lies on the boundary between the Pacific and North American Plates which are moving past each other at a rate of about 6 cm a year (Figure 1.10). The whole of southern California is criss-crossed by **fault lines** caused by the movement of these two plates. The focus of the Los Angeles earthquake was traced to 5 km below the surface on the Oakridge fault.

Forty-five people were killed and nearly 2000 were injured. At least 650 000 buildings were damaged and thousands of people were left without gas, water or electricity (Figure 1.11). Broken gas mains started over 100 fires. Communications were affected as well – telephone wires were cut and many roads and bridges were destroyed. The repair bill was estimated at $15 billion.

Word box

aftershocks The smaller tremors which happen after the main earthquake

fault line A crack in the crust along which the rocks move

The emergency services were quick to respond to the situation and 2000 fire-fighters were brought in from the rest of California to search through the rubble for survivors. The police set up a dusk-to-dawn curfew to stop looting. The Red Cross set up 22 camps to help house the 20 000 people who had to leave their homes (Figure 1.12). The American Government gave $100 million of government money to the relief fund and promised long-term help such as low interest loans. Life was back to normal for most people in only a few weeks, although it is likely to take nearly two years for all of the repairs to be completed.

The damage would have been greater but the building controls in Los Angeles are very strict. However, no building can stand up to a really big earthquake; for example, in the 1970s, $1.5 billion was spent on strengthening roads but many of them collapsed in this earthquake.

Figure 1.12 Temporary shelter, Northridge, Los Angeles after 1994 earthquake

The previous two big earthquakes in Los Angeles were in 1971 and 1934 and it is certain that there will be more in the future. However, most of the eight million inhabitants are prepared to take the risk of 'being around for the big one' rather than leaving the city.

Figure 1.13 An eye witness account of the 1994 Los Angeles earthquake

The first sign of the earthquake was my cat bolting from the bed. The tinkling of glasses like fingertips against window panes gave way to the rattling of pots and pans in kitchens for several blocks. After two years in Los Angeles I am used to running towards a door jamb, the safest place to stand, only to feel the floor settle before I make it. This time I knew there was trouble when I scarcely reached the door without being thrown to the ground. The bedroom was rocking like a boat tossed on a stormy ocean. Car alarms screeched in the street below. The floor was rolling in waves. Only a day earlier I had convinced myself that nothing could beat the 17°C January days of southern California. Now I was not so sure. Gripping the frame of the door, I stared at the hallway ceiling and saw it bulging towards me. The rest of the 45 seconds or so of the earthquake are a blur.

The Times, 18 January 1994

QUESTION BOX

1 For the 1994 Los Angeles earthquake, write one or two sentences about each of the following: When did it happen? Where did it happen? Why did it happen? What damage did it do? How did the city try to cope?

A Describe and explain the causes and consequences of the 1994 Los Angeles earthquake. How successfully did the people of Los Angeles cope with this earthquake?

CASE STUDY

The Afghan Earthquakes, 1998

In the first six months of 1998 two powerful earthquakes hit an **inaccessible** mountainous region in the north of Afghanistan (Figure 1.14). The first was on Wednesday, 4 February and measured 6.1 on the Richter scale. The second was on Saturday, 30 May and measured 7.1 on the Richter scale.

Earthquakes are common in Afghanistan. It lies on a plate boundary where the Arabian and Indo-Australian plates are moving into the Eurasian plate (Figure 1.15). The Himalayas are the result of this collision.

Figure 1.14 Location map; Afghanistan earthquake, 1998

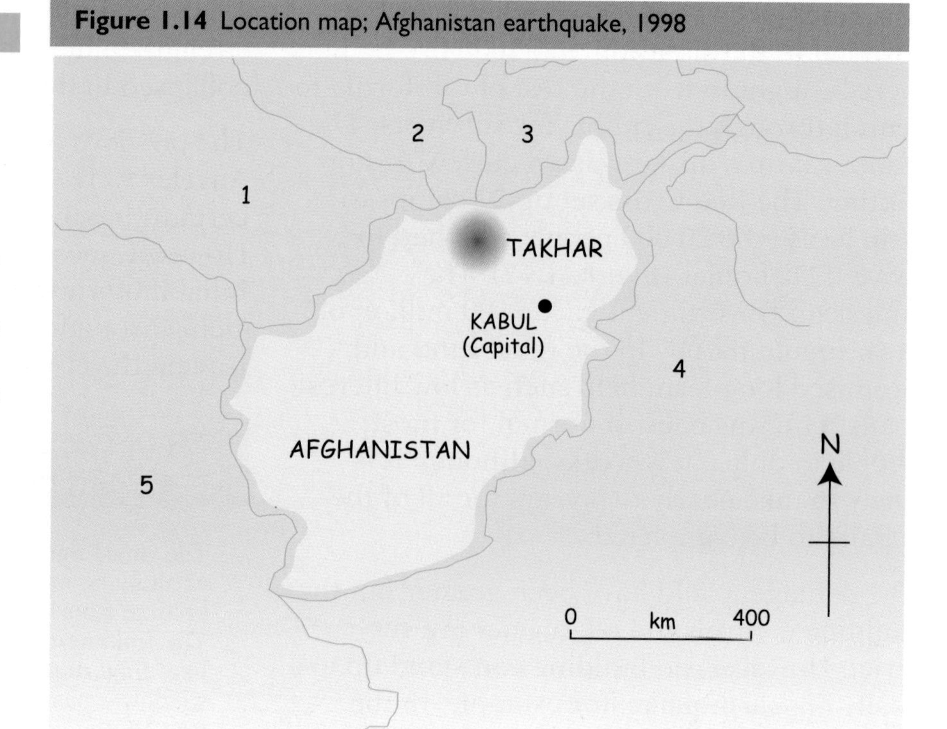

area affected by earthquakes

1 Turkmenistan
2 Uzbekistan
3 Tajikstan
4 Pakistan
5 Iran

Figure 1.15 Plate boundaries in the Afghanistan region

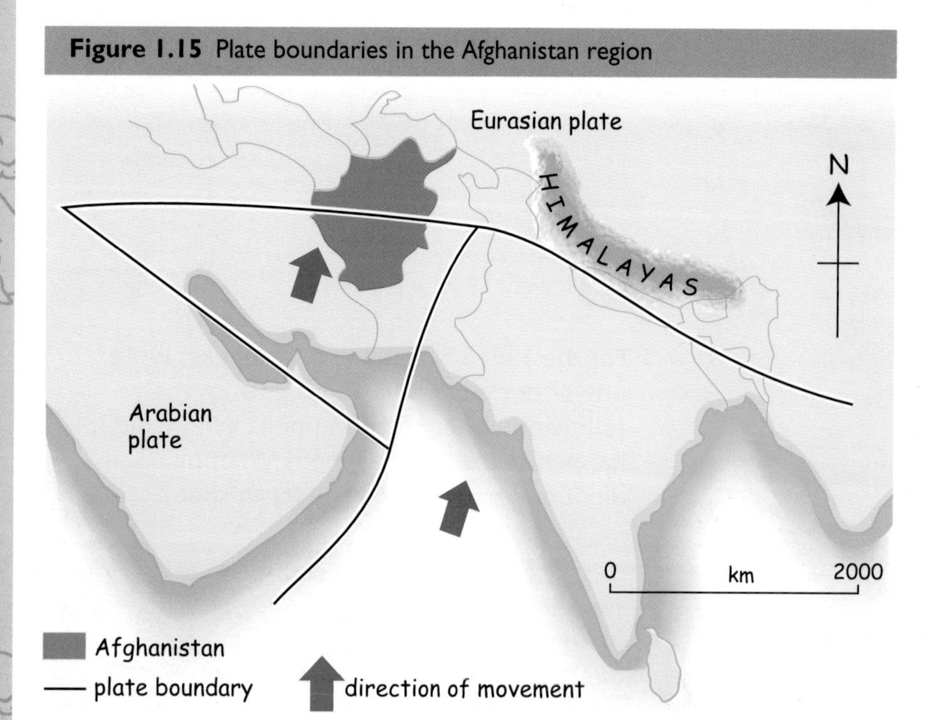

It was several days before the outside world heard of the disaster caused by the February earthquake. Getting **foreign aid** through was difficult for a number of reasons:

- the roads are very poor and they were blocked by landslides caused by the earthquake
- the weather was bad
- this is a part of the country where there is fighting between the government and rebel forces

In total, over 2300 people were killed and 8000 homes were destroyed.

The second earthquake killed over 4000 people. The death toll would have been higher but it happened during the day when many people were outside. Rescue operations were in full swing within 48 hours because there was food and equipment in the region left over from February. However, it still took five days to reach some of the remote villages and disease quickly became a problem because wells, springs and water canals had been damaged so there was very little fresh water.

Word box

inaccessible Difficult to get to

foreign aid Help (e.g. money, food, medical supplies) given by one country to another

QUESTION BOX

1 Make a copy of Figure 1.14. Label onto it the main facts about the 1998 Afghan earthquakes: when and where they happened; their strength; and the damage they caused.

2 Describe the scene in Figure 1.16. What building materials can you identify? What damage has been done? What do you think the people in the photograph might be thinking?

A Explain why Afghanistan has so many earthquakes.

B Give three reasons why these earthquakes caused more damage than the 1994 Los Angeles earthquake.

C Why was it harder for the Afghan government to deal with these earthquakes than for the American government to deal with the 1994 Los Angeles earthquake?

Figure 1.16 Image taken from television footage of earthquake damage in Afghanistan, 1998

unit 2

Changing Landscapes

Key questions

- How does water move through the atmosphere and environment?
- How does a river landscape change from its source to the sea?
- How does a river shape the landscape?
- How have rivers been used and abused by people?
- What causes flooding?
- How can flooding be reduced?

Water

Water makes up 70 per cent of the surface of our planet. Over 99 per cent of water is in oceans and seas or ice caps and glaciers (Figure 2.1). Water is essential to life; 66 per cent of your body is water. In many parts of the world there is a water shortage but in more developed countries this resource is taken for granted.

Figure 2.1 Global water distribution

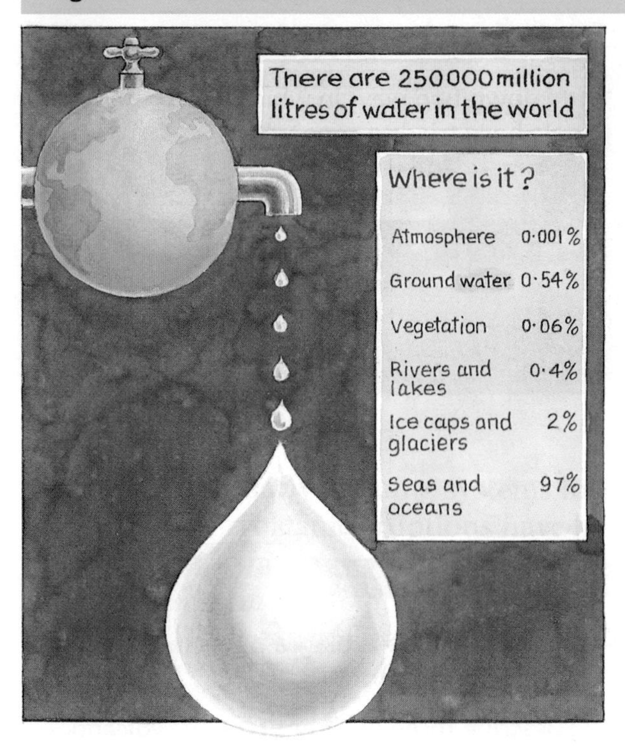

The water cycle

Water changes as it moves through the **water cycle** (Figure 2.2). The sun's energy heats the water and it is evaporated from the oceans and the earth's surface as water vapour. The warm moist air carrying the vapour rises upwards. In the upper atmosphere the water vapour is cooled and condenses. The condensation forms tiny water droplets around particles of dust or ice in the atmosphere. The droplets form snow flakes. When these become too large they fall as **precipitation**, becoming sleet and then rain in the warmer lower atmosphere.

Movement through a drainage basin

As rainwater reaches the ground it flows over the surface as streams or rivers or soaks into the ground. Rivers drain large areas of the land which are called **drainage basins**. The ridge of higher land between drainage basins forms a dividing line called the **watershed** (Figure 2.3).

The water moves through a drainage basin as a river or as underground flow (Figure 2.4). Some of the rainwater falls on plants, trees or buildings. This is evaporated along with water from the ground and from rivers. Vegetation releases water back into the atmosphere by **evapotranspiration**. Some water sinks into the soil and slowly moves through the rock as ground water flow. A river will only flow over land where the rock is **impermeable** or where the **water table** occurs at the surface.

In total, over 2300 people were killed and 8000 homes were destroyed.

The second earthquake killed over 4000 people. The death toll would have been higher but it happened during the day when many people were outside. Rescue operations were in full swing within 48 hours because there was food and equipment in the region left over from February. However, it still took five days to reach some of the remote villages and disease quickly became a problem because wells, springs and water canals had been damaged so there was very little fresh water.

QUESTION BOX

1. Make a copy of Figure 1.14. Label onto it the main facts about the 1998 Afghan earthquakes: when and where they happened; their strength; and the damage they caused.
2. Describe the scene in Figure 1.16. What building materials can you identify? What damage has been done? What do you think the people in the photograph might be thinking?

A Explain why Afghanistan has so many earthquakes.

B Give three reasons why these earthquakes caused more damage than the 1994 Los Angeles earthquake.

C Why was it harder for the Afghan government to deal with these earthquakes than for the American government to deal with the 1994 Los Angeles earthquake?

Word box

inaccessible Difficult to get to
foreign aid Help (e.g. money, food, medical supplies) given by one country to another

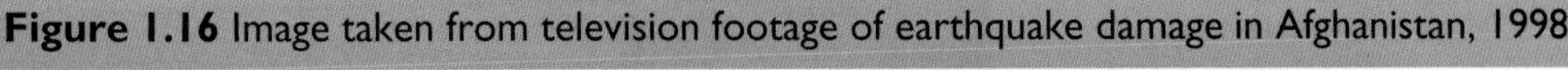

Figure 1.16 Image taken from television footage of earthquake damage in Afghanistan, 1998

Managing hazards: earthquakes and volcanoes

We cannot stop earthquakes and volcanic eruptions. However, sometimes it is possible to change the course of a lava flow but only if it is moving very slowly. For example, in 1983 the Italians managed to steer a lava flow from Mount Etna, on the island of Sicily, away from some houses by building a rock barrier; but it took four months and cost £3 million (Figure 1.17).

Figure 1.17 Lava flow from Mount Etna causing damage to crops

Scientists are getting better at predicting when earthquakes and volcanic eruptions are going to happen. **Seismographs** can be used to record earth tremors (Figure 1.18); lasers can be used to measure tiny amounts of movement on either side of a fault line; and the level of water in wells can be recorded. There have been some successes but not very many; for example, Chinese scientists predicted an earthquake measuring 7.3 on the Richter scale in Haicheng in 1975 and saved thousands of lives by evacuating the city before it happened. However, the same scientists failed to predict an earthquake in Tangshan a year later which caused 240 000 deaths.

Figure 1.18 Seismograph

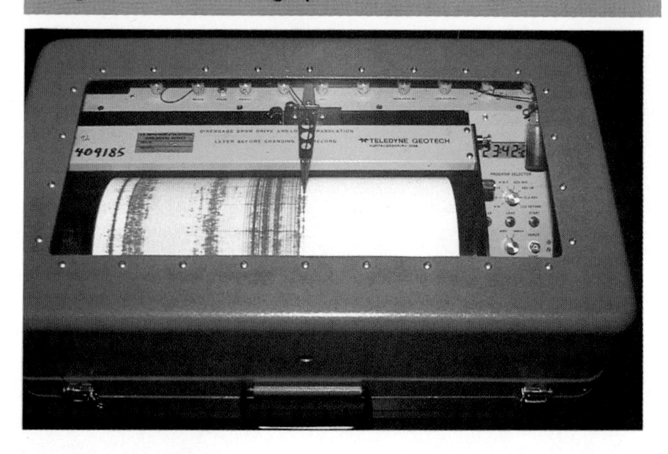

Figure 1.19 Mount Aso with crater, Japan

Early warning systems for volcanic eruptions have had more success. For example, in Japan pressure pads have been placed around some volcanoes (Figure 1.19). When the lava at the start of an eruption lands on them, alarms in nearby villages are set off and video cameras are switched on so that the eruption can be monitored.

Word box

seismograph A piece of equipment which is used to measure earth tremors

It is also possible to build safer houses. This is easier to do in rich countries than in poor ones but there are ways of building safer houses even with cheap materials; for example, planks of timber have been used to 'strap' stone, rubble or clay houses together for thousands of years in Turkey (Figure 1.20).

Figure 1.20 Safe houses save lives

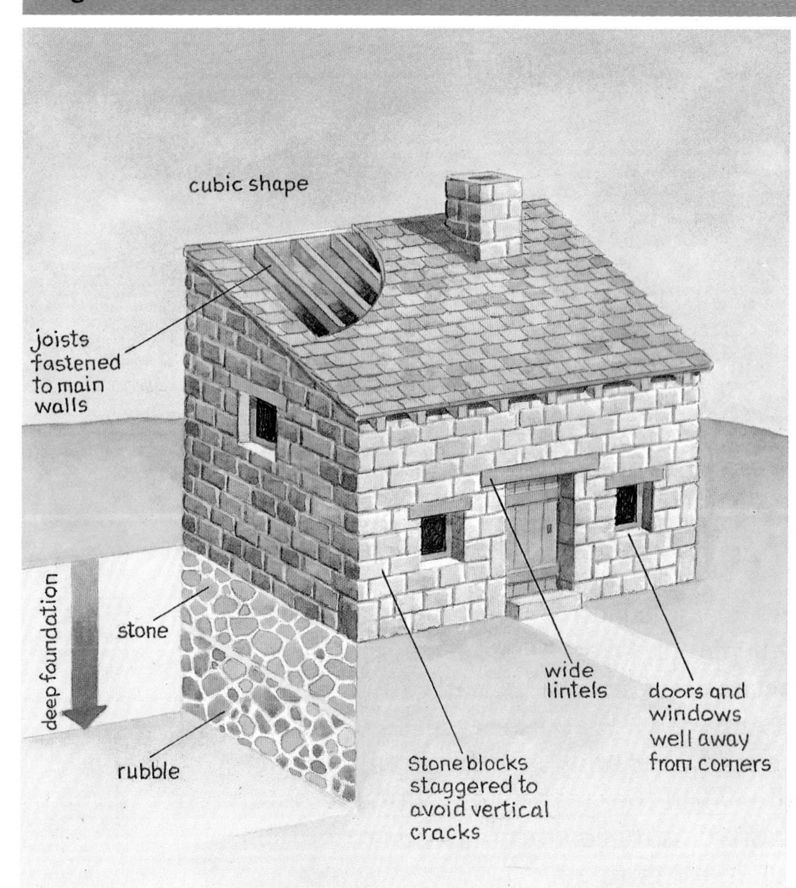

QUESTION BOX

1 Write down two ways in which people have tried to cope with earthquakes and volcanoes.

A Explain why a rich country like the USA stands more chance of predicting an earthquake than a poor country like India?

SUMMARY QUESTIONS

1) Draw and label a diagram to show the main layers of the earth.

2) Describe three different types of volcano.

3) Mention some of the good and bad things about volcanoes.

4) Compare the 1994 Los Angeles earthquake and the 1998 Afghanistan earthquake using these sub-headings: strength; deaths and injuries; damage; help from the government; and help from other countries.

A) Write down three reasons why some earthquakes cause more damage than others.

B) Why did Los Angeles cope better with the 1994 earthquake than Afghanistan coped with the 1998 earthquake?

C) What can we do to lessen the threat from earthquakes and volcanoes?

D) Atlases and encyclopaedias usually have a list of major earthquakes. Find one of these lists and mark the most recent 15 or 20 earthquakes onto a world map. Label onto the map the damage caused by each of the earthquakes. Then, describe and explain any pattern your map shows.

unit 2

Changing Landscapes

Key questions

- How does water move through the atmosphere and environment?
- How does a river landscape change from its source to the sea?
- How does a river shape the landscape?
- How have rivers been used and abused by people?
- What causes flooding?
- How can flooding be reduced?

Water

Water makes up 70 per cent of the surface of our planet. Over 99 per cent of water is in oceans and seas or ice caps and glaciers (Figure 2.1). Water is essential to life; 66 per cent of your body is water. In many parts of the world there is a water shortage but in more developed countries this resource is taken for granted.

Figure 2.1 Global water distribution

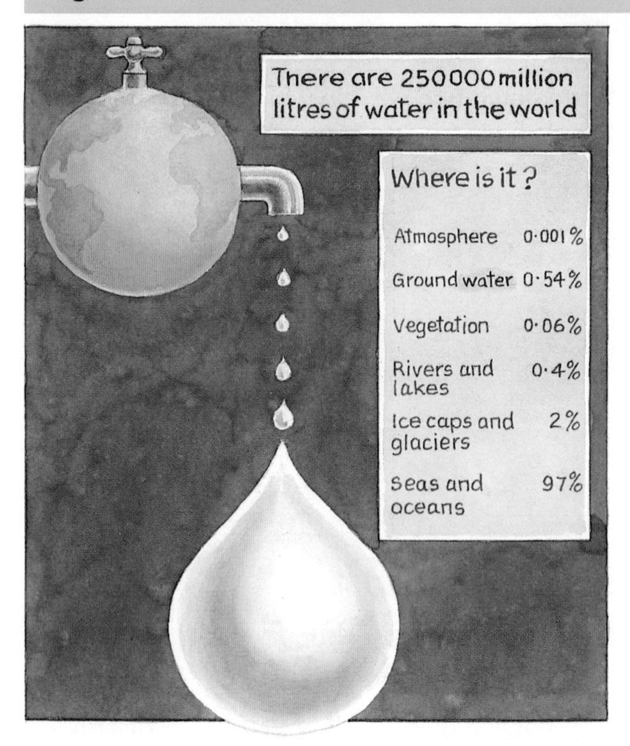

The water cycle

Water changes as it moves through the **water cycle** (Figure 2.2). The sun's energy heats the water and it is evaporated from the oceans and the earth's surface as water vapour. The warm moist air carrying the vapour rises upwards. In the upper atmosphere the water vapour is cooled and condenses. The condensation forms tiny water droplets around particles of dust or ice in the atmosphere. The droplets form snow flakes. When these become too large they fall as **precipitation**, becoming sleet and then rain in the warmer lower atmosphere.

Movement through a drainage basin

As rainwater reaches the ground it flows over the surface as streams or rivers or soaks into the ground. Rivers drain large areas of the land which are called **drainage basins**. The ridge of higher land between drainage basins forms a dividing line called the **watershed** (Figure 2.3).

The water moves through a drainage basin as a river or as underground flow (Figure 2.4). Some of the rainwater falls on plants, trees or buildings. This is evaporated along with water from the ground and from rivers. Vegetation releases water back into the atmosphere by **evapotranspiration**. Some water sinks into the soil and slowly moves through the rock as ground water flow. A river will only flow over land where the rock is **impermeable** or where the **water table** occurs at the surface.

Figure 2.2 The water cycle

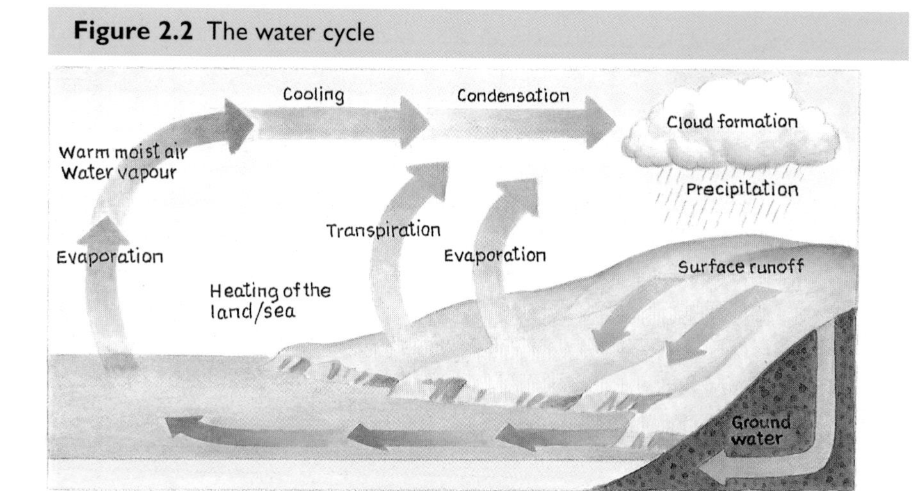

The flow or amount of water moving through a drainage basin depends on several factors. These include:

- geology – the type of rock found in the drainage basin (permeable rocks will have less surface water);
- seasonal climatic conditions – the higher evaporation rates in summer reduce the amount of surface water;
- infiltration rates – the amount of water which soaks into the ground will vary according to soil type;
- vegetation cover and land use.

Word box

water cycle The movement of water through the land, sea and atmosphere
precipitation Wet weather including rain, sleet, hail and snow
drainage basin Also called the catchment area, an area of land drained by a river system
watershed A boundary between two drainage basins
evapotranspiration The movement of water from plants back into the atmosphere
impermeable A rock which fluids (e.g. water) cannot pass through
water table the level below which the ground is saturated and where water can be found

Figure 2.3 Side and plan view of a drainage basin and watershed

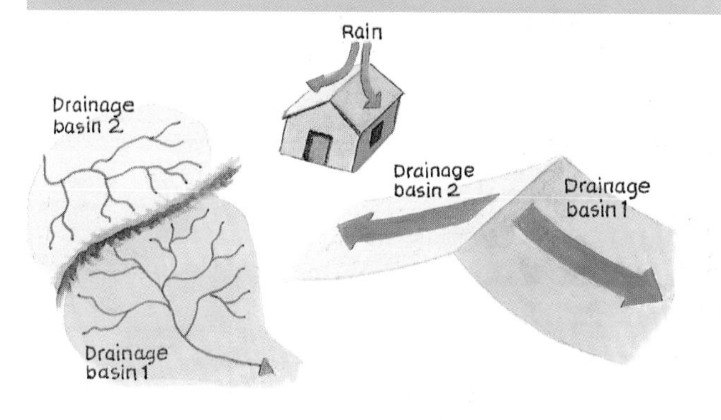

Figure 2.4 Movement of water through a drainage basin

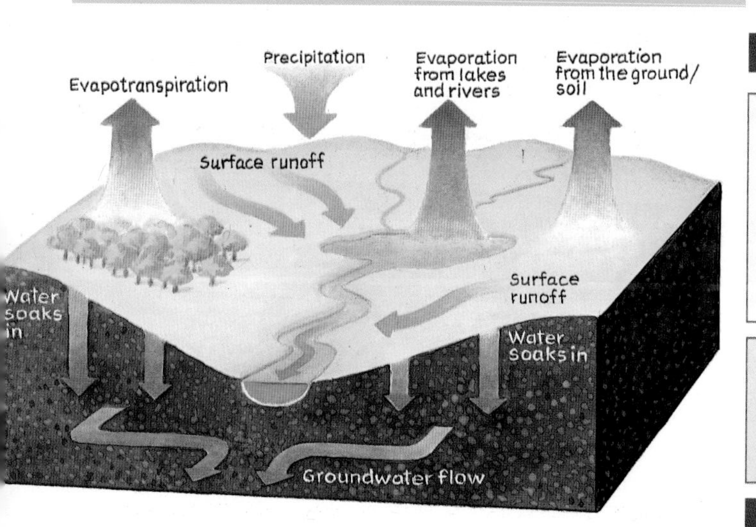

QUESTION BOX

1 Draw a labelled diagram to describe the water cycle.

2 Describe the global distribution of water.

3 State **two** ways water can move out of a drainage basin.

A Explain how the water cycle works.

B Draw a flow chart to show the movement of water through a drainage basin.

Extension task: see worksheet 2A/B

River landscapes

Figure 2.5 Contour map and long profile of a river

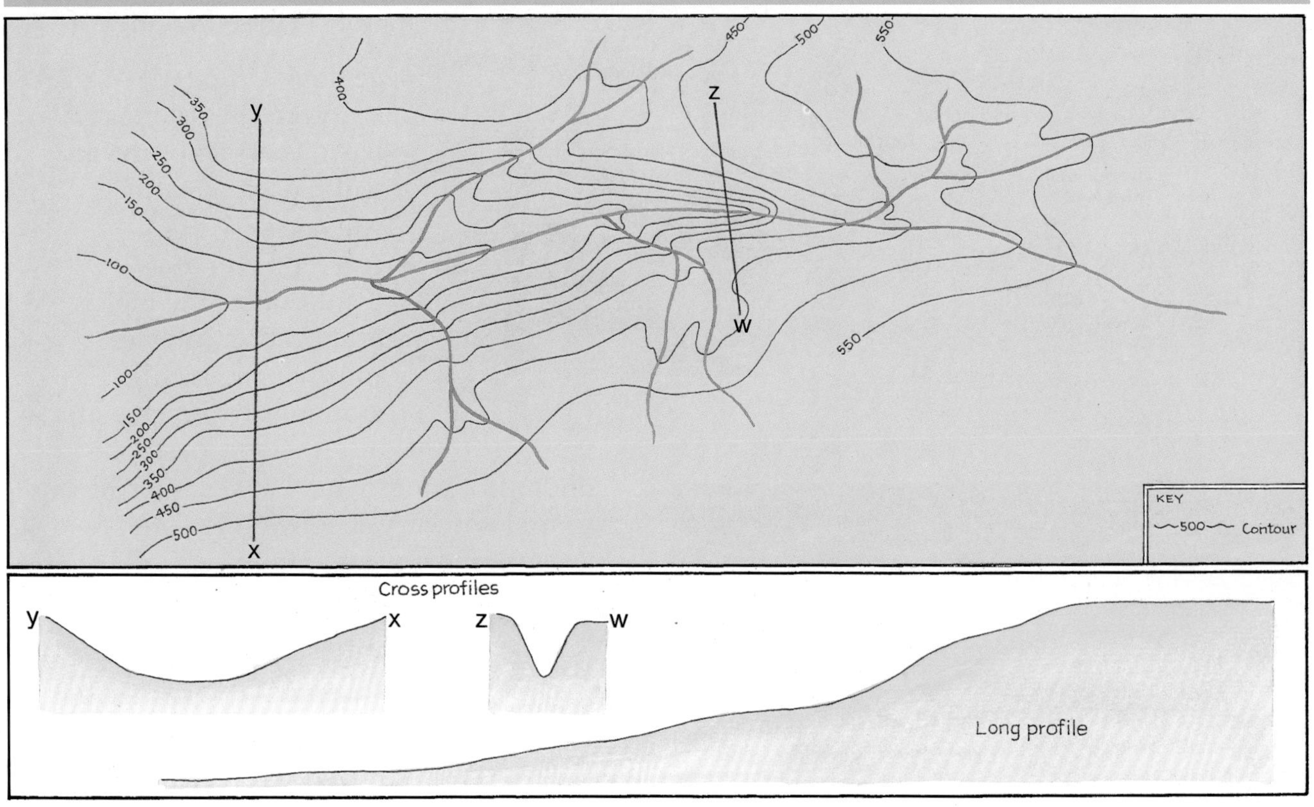

A river **erodes** the landscape, transports sediment and then deposits the sediment along its course. To change the landscape a river uses energy. Figure 2.5 shows the **long profile** of a river. The energy a river has depends on how fast it is flowing and the discharge. The **gradient** of the river course and the river discharge affects the velocity (speed) of the river. The faster the river flows and the larger the flow, the more energy a river has to erode and transport sediment. Sediment is deposited when the river loses energy. The appearance of a river valley changes from a river's source to its mouth (Figures 2.6 and 2.7).

Figure 2.6 Profile of a river

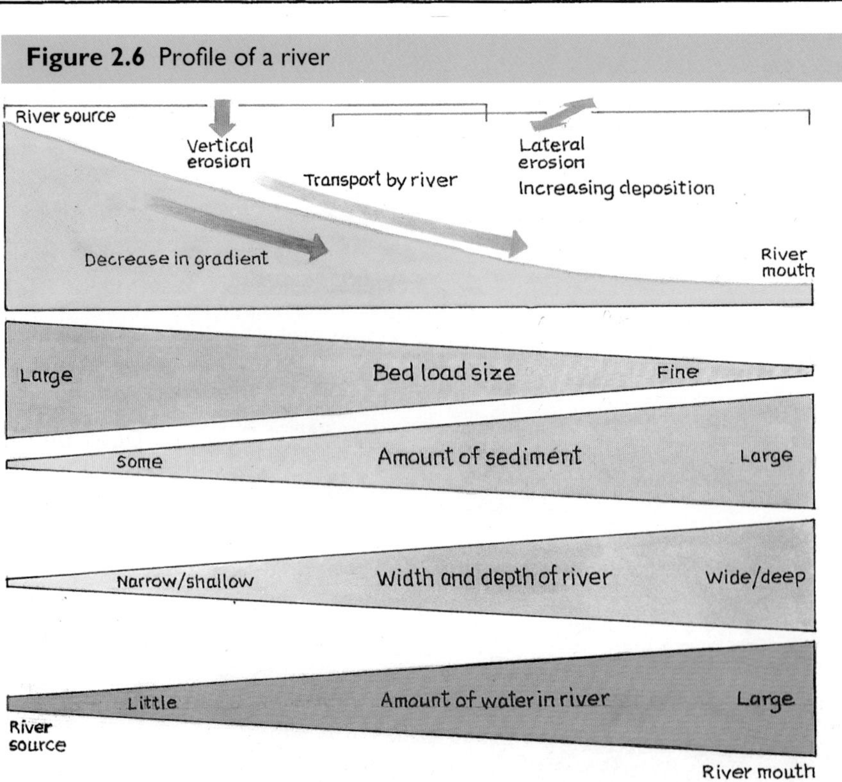

At its source a river is small and fast flowing in a narrow valley with a steep gradient. The V-shaped valley shows that the stream is eroding downwards, cutting a narrow slot in the land. Although the river discharge is small, the fast flow enables the river to erode rapidly into the land. The valley develops **interlocking spurs** with rapids and waterfalls along the river course.

In its lowest course the gradient is flatter. Here the river is eroding laterally (sideways) forming a wide valley floor. The river is flowing more slowly and features of deposition are common, but it has a greater volume of water which still enables it to erode the landscape. The river develops large river bends called meanders. Over time these meanders move slowly across and down the valley. As they move across they widen the valley. The flat valley floor is called the **flood plain**, where the river will flood in wet weather.

Figure 2.7 Comparing features and process of the upper and lower courses of a river

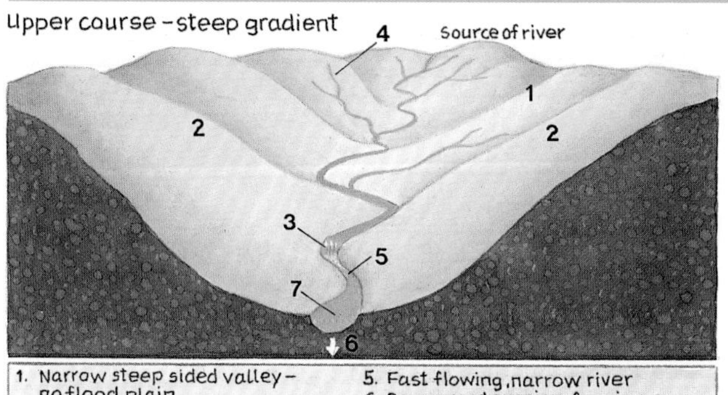

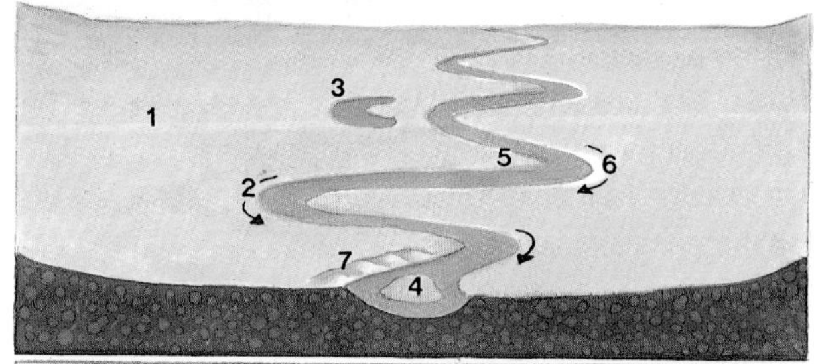

Figure 2.8 a) V-shaped valley b) Flood plain

QUESTION BOX

1 State the **three** ways in which a river shapes the landscape.

2 Describe the characteristics of:

a) the upper part of a river valley;

b) the lower part of a river valley.

A Using Figure 2.6, explain the link between the flow of a river and how a river shapes the landscape.

B Explain why the cross profile of the upper part of the river is different to the cross profile of the lower part of the river.

Word box

erodes or erosion The wearing away of the land by wind, water, gravity or ice

long profile A chart showing the gradient of a river, drawn from contours

gradient How steep the river course is

interlocking spurs Steep ridges of land which jut alternately into a river valley, around which a river runs its course

flood plain A low flat area either side of a river in its lowest course

River processes

A river erodes the landscape by:

- corrasion, where stones carried along by the flow of the river knock and rub against the river bed and bank and so wear it away;
- corrosion, where the minerals which make up the river bed and bank are dissolved by the water;
- hydraulic action, where water or air trapped in the cracks in the river bed or bank are pressurised by the flow of the river. This pressure weakens the rock and causes it to disintegrate (Figure 2.9).

Figure 2.9 How a river erodes

Figure 2.10 Formation of a waterfall

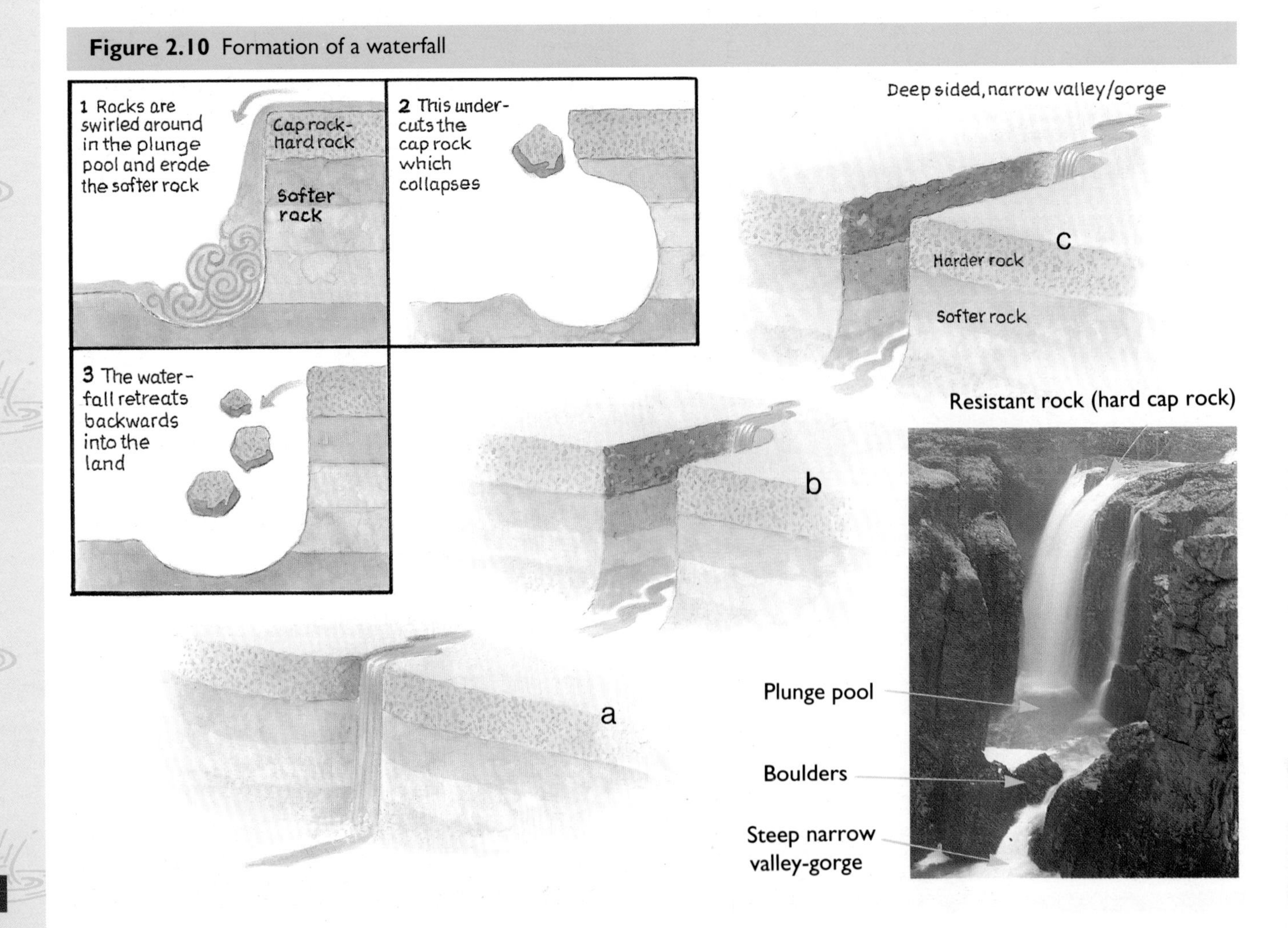

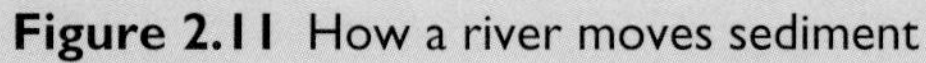

Figure 2.11 How a river moves sediment

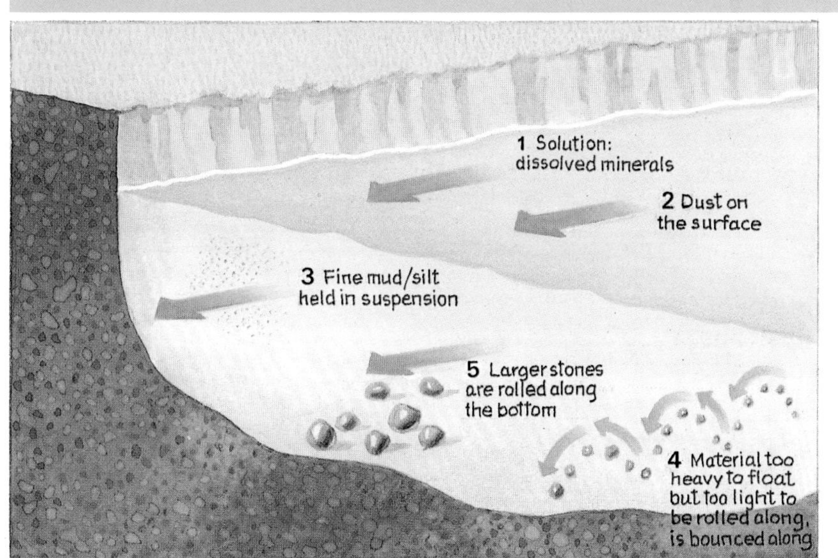

Word box

gorge A narrow valley often formed in resistant rock

bedload The material transported by a river, usually sands and pebbles

attrition The wearing down and smoothing of sediment by particles, e.g. pebbles, knocking against each other

Softer rocks like clay and shale are eroded at a quicker rate than harder and resistant rocks like granite. This process can be seen in the formation of a waterfall/**gorge** (Figure 2.10).

A river can be described as a conveyor belt which moves sediment along the river course down to its mouth where the material is deposited (Figure 2.11). The size of sediment moved by a river decreases downstream.

At the source the **bedload** is made up of big, irregular shaped rocks. Downstream the bedload is finer and the stones worn round and smooth by **attrition**. At its mouth the deposits are a fine mud. As a river enters the sea it loses all of its energy and deposits any sediment it is carrying. This sediment may build up to form mudflats, e.g. the Thames Estuary (Figure 2.12).

Figure 2.12 Deposition of sediment at an estuary
Inset photo of the Thames Estuary

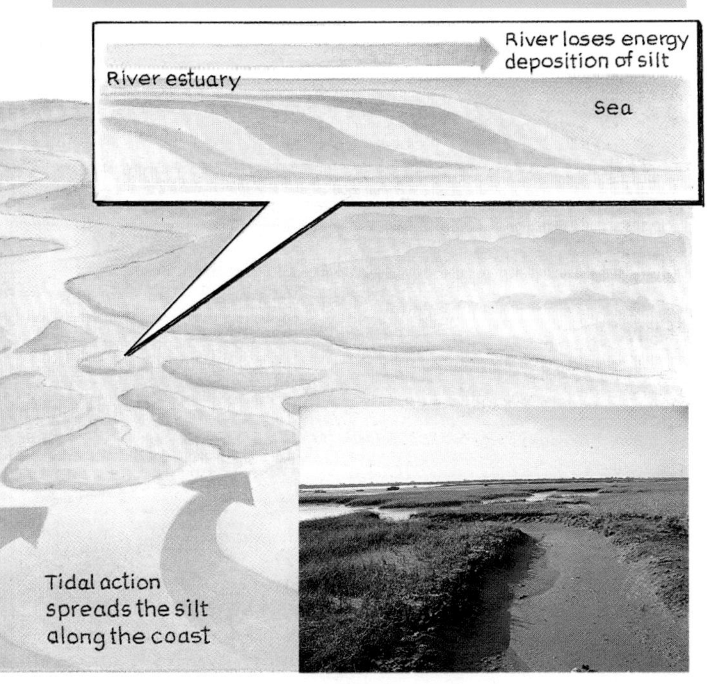

QUESTION BOX

1. List **five** ways a river erodes its course.
2. Describe **two** ways a river moves sediment along its course.
3. Use labelled drawings to show what processes happen at a waterfall.

A Explain how a river can erode its course.

B Explain why a river can be described as a conveyor belt?

C Choose and explain the processes which produced a river feature formed by erosion and one formed by deposition.

Extension task: see worksheet 2C/D

USE AND ABUSE OF RIVERS

The River Yangtze

Rivers are an important part of our landscape and also affect our lives in various ways. They are often used by people for transport routes and may provide a location for economic development, for example the Yangtze River in China. Yet many rivers pose a threat to the people who live by and depend upon them (Figure 2.13).

As a country invests in its economic development the river basin can become vulnerable. The 1998 floods in China have been blamed on **deforestation** in catchment areas which has caused flooding by rapid run off. LEDCs have to repay debts and invest in exports. This limits a government's ability to maintain the old river defences like **levées** (Figure 2.14). People's ability to survive severe flooding is related to their poverty and the availability of healthcare and welfare programmes to help them cope following a disaster.

Figure 2.13 Flood risk in river basins

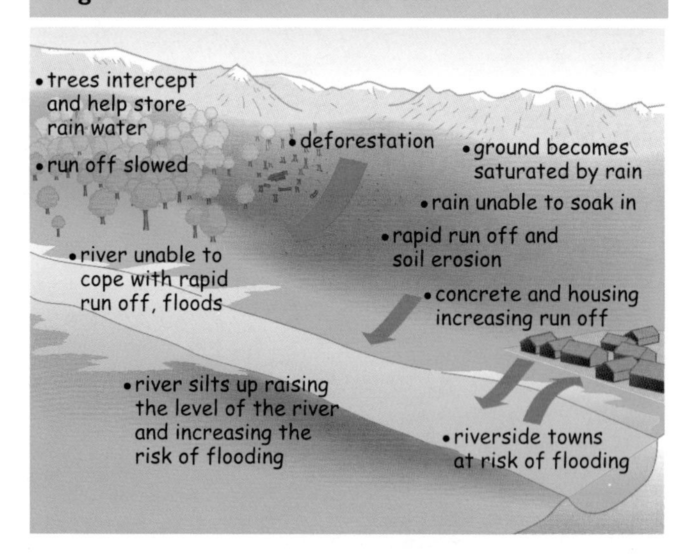

Figure 2.14 River levées

Yangtze Floods 1998

The Yangtze is the world's third longest river. It flows 4000 miles from Tibet to its mouth at Shanghai. Over 350 000 people have died from flooding since 1931. In 1998, 233 million people were affected by the flood

Figure 2.15 Location of the Yangtse river

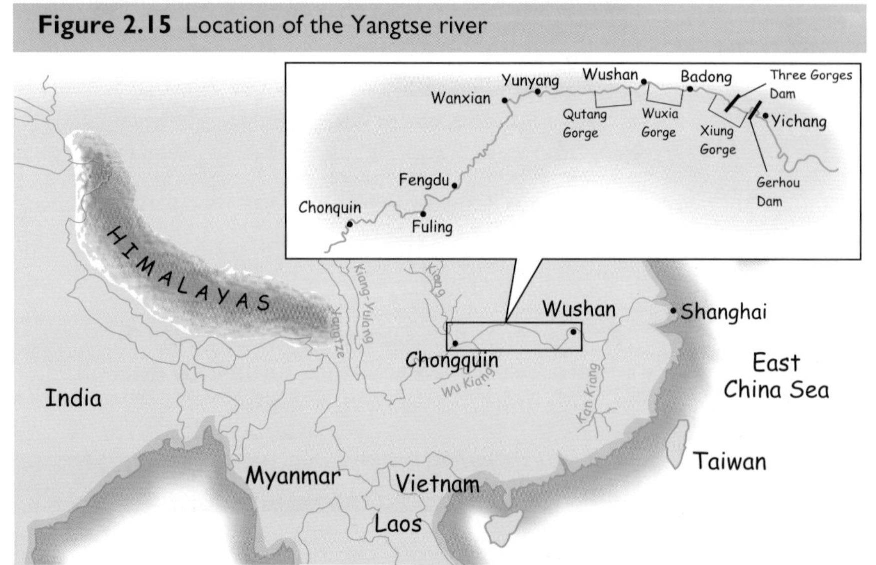

Word box

levée Originally a natural river bank formed by deposits of sediment following flooding, though many have been artificially raised to form river defences

deforestation Completely clearing an area of trees

HEP Hydro electric power, electricity produced by water driven generators

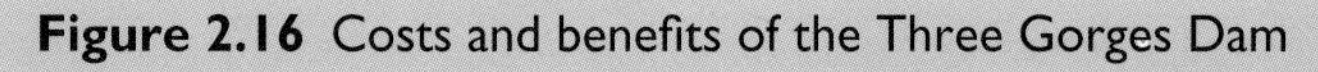

Figure 2.16 Costs and benefits of the Three Gorges Dam

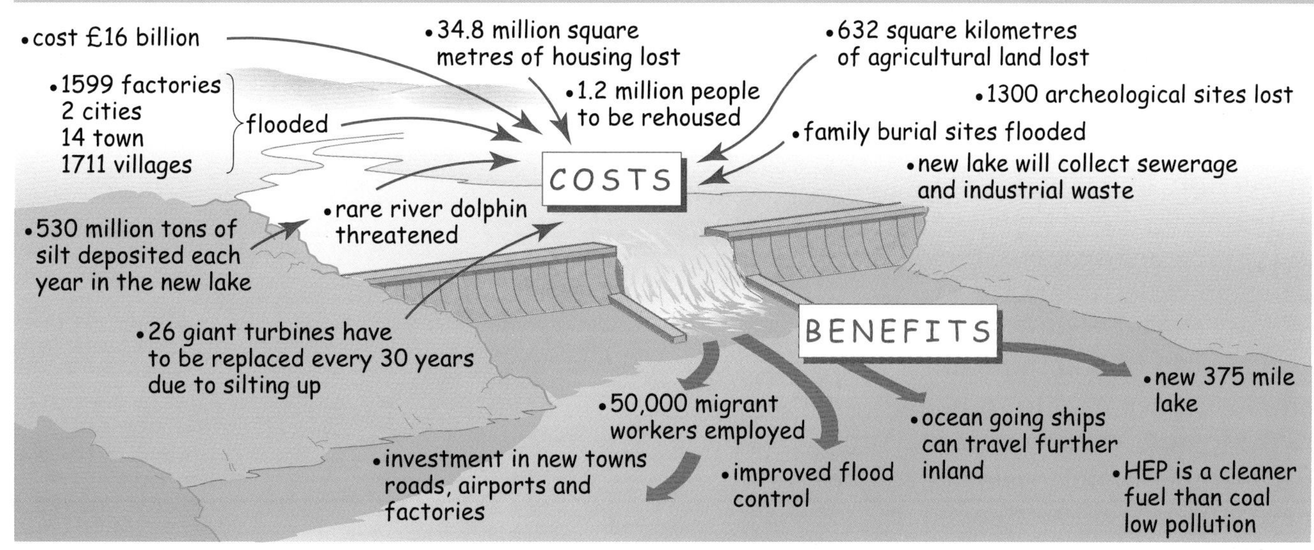

surges in the Yangtze river system which flooded central China. Over 3500 people died, 15 million people were displaced and 5 million people lost their homes. 270 000 hectares of crops were destroyed reducing the grain harvest by 11 million tonnes. The floods cost £20 billion and it will take years to repair the damage (Figure 2.15). Illness will claim more lives.

The Three Gorges Dam

The dam was planned to provide cheap **HEP** for China's industry and encourage economic growth. It also aimed to control the Yangtze floods. The lake created will be 375 miles long. About 40% of the £16 billion costs will go to build new cities for those displaced. There is concern about the environmental impact of the dam and how long it can survive the Yangtze's annual sediment load of 530 million tonnes of silt being deposited in the lake (Figure 2.16).

Figure 2.17 Flooding of the Yangtze River, 1999

QUESTION BOX

1 Look at Figure 2.16 and an atlas.

a) Draw a map of the Yangtze River and label its main tributaries.

b) Write a description to locate the Yangtze River.

2 What caused the Yangtze to flood?

3 How did the 1998 flooding affect China?

A Suggest why people in an LEDC find it difficult to cope with floods.

B Write a summary of the Three Gorges Dam, include:

a) a map locating the Three Gorges Dam;

b) comment on the advantages for China;

c) any problems it may cause;

d) your ideas on how successful you think this scheme will be.

River Management

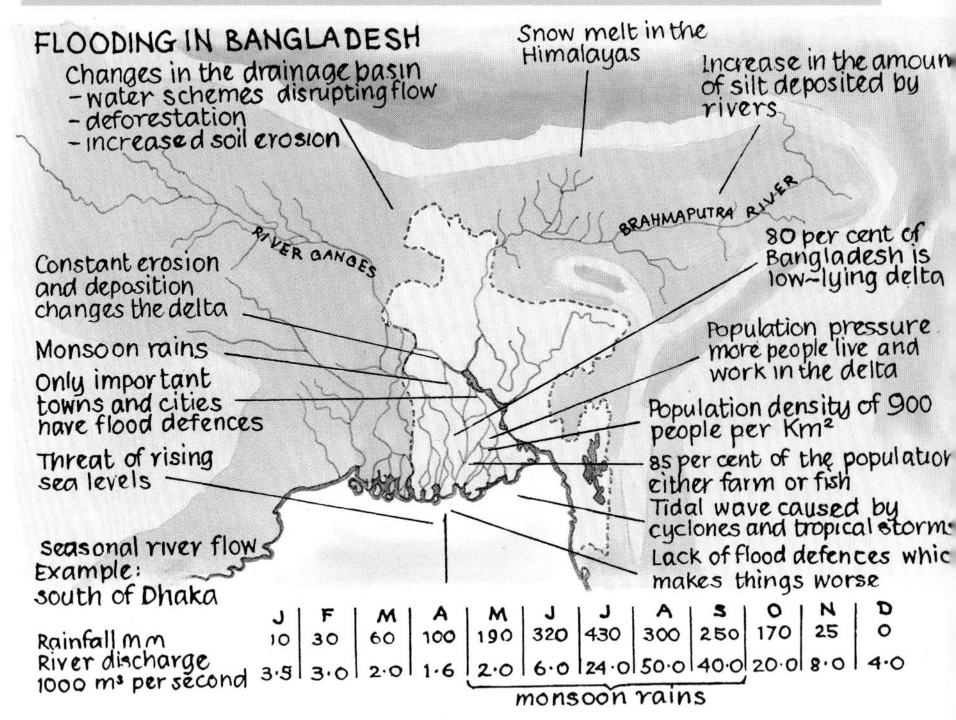

	J	F	M	A	M	J	J	A	S	O	N	D
Rainfall mm	10	30	60	100	190	320	430	300	250	170	25	0
River discharge 1000 m³ per second	3·5	3·0	2·0	1·6	2·0	6·0	24·0	50·0	40·0	20·0	8·0	4·0

Figure 2.18 Flooding in Bangladesh

Not even the USA, one of the world's richest and most technically advanced countries could stop the Mississippi flooding in 1993. Although the flood defences did protect many areas, the damage was estimated at $10 billion. The government provided $3 billion in relief, and insurance paid for much of the losses. Many countries, for example Bangladesh, have far less money to spend on protecting their people and buildings, so the effects of the flooding are greater. There are many causes of flooding in Bangladesh (Figure 2.18). Some floods are beneficial to farmers whose fields are then covered with nutrient rich silt, however, flooding can also bring disaster. In 1998 prolonged flooding killed a thousand people, 24 million people lost their homes, 700 000 hectares of crops were lost, 2.5 million farmers were affected and schools and hospitals were destroyed as 34 000 square miles were flooded. Disease and hunger will kill many more in the contaminated region. It will take many years to replace the fresh water wells and for those made destitute to rebuild their lives.

Bangladesh has proposed a major scheme to reduce the effects of flooding (Figure 2.18). With a national debt of $16 billion and an economy based on industry damaged by the floods, can Bangladesh afford the scheme?

QUESTION BOX

1 Suggest why an MEDC like the USA can cope with flooding better than an LEDC like Bangladesh.

2 Write a short account for a newspaper which is reporting the main facts about the floods in Bangladesh.

A Write an imaginary interview with a victim of the 1998 Bangladesh flood. Think of **ten** questions to ask and the likely answer to each.

B Use Figure 2.21 to suggest ways the threat of flooding be reduced in Bangladesh.

a) Which would have little effect, say why.

b) Which would have some effect, say why.

c) Suggest some other ideas.

Figure 2.19 Bangladesh flood protection

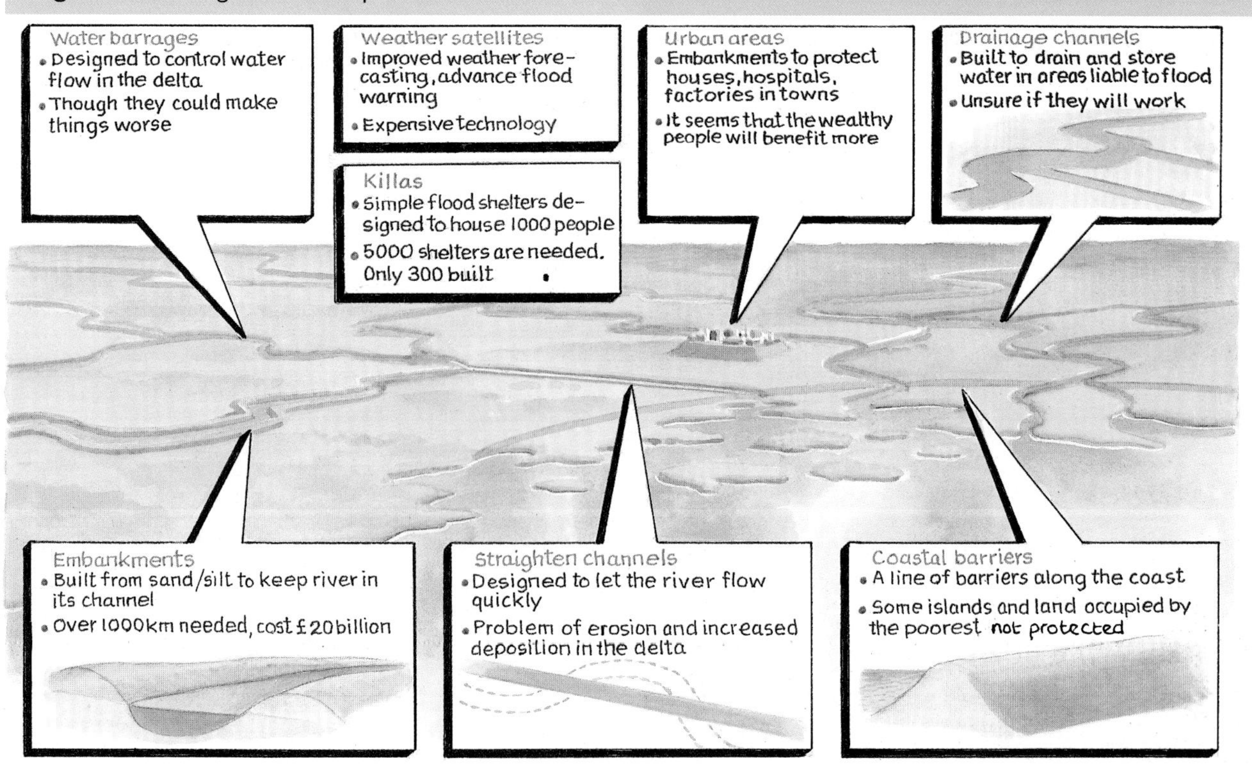

SUMMARY QUESTIONS

1 Use these words to write a paragraph to explain how water moves through the atmosphere:

evaporation; condensation; precipitation; surface run off; groundwater

2 Use these words to write a paragraph to explain how water moves through a drainage basin:

watershed; drainage basin; precipitation; surface run off; groundwater flow; evaporation; evapo-transpiration; groundwater flow

3 Draw a sketch to show an upper part of a river and a lower part of a river.

4 Explain, using diagrams, one way a river erodes its course.

5 State **three** ways in which rivers have been used and abused by people?

6 Suggest a reason why people live in areas likely to flood.

A Which factors can affect the movement of water through a drainage basin?

B Explain why the upper valley river landscape is different from the lower part of the course.

C How does a river shape the landscape?

D Do you think that building the Three Gorges Dam will benefit the local population? Explain your answer.

E Explain how human activity can cause flooding?

unit 3

The Weather Machine

Key questions

- What is the difference between weather and climate?
- How and why does the climate of the British Isles vary from place to place?
- How do weather and climate affect human activity?

Weather vs climate

Figure 3.1 gives the weather reports for a week in September in Skegness. Each day was a little bit different. This is what we mean by **weather**: the day to day changes in sun, rain and clouds.

Figure 3.2 gives the temperature and rainfall statistics for Skegness for a whole year. It shows that Skegness has warm, quite dry summers and cool, quite dry winters: this is Skegness's **climate** – its 'average' weather.

Figure 3.1 Weather reports for Skegness

Day	Highest temperature°C	Lowest temperature°C	Rain mm	During the day
3 September 94	18	9	–	sunny
4 September 94	19	11	1.0	sunny
5 September 94	18	11	–	bright
6 September 94	16	10	–	cloudy
7 September 94	17	9	3.0	sunny
8 September 94	19	10	3.0	sunny

Figure 3.2 Climate statistics for Skegness

Month	J	F	M	A	M	J	J	A	S	O	N	D
Temp °C	4	4	6	8	11	14	16	16	14	11	7	5
Rain mm	55	40	35	40	45	45	55	60	50	50	60	45

Word box

weather Day to day changes in rainfall and temperature
climate What the weather is like from season to season
range of temperature The difference between the highest and lowest temperature – take away the lowest from the highest
total yearly rainfall A whole year's (annual) rainfall, adding together the rain for each month

The climate of the British Isles

Figure 3.3 shows the pattern of rainfall in the British Isles and gives you information about winter and summer temperatures. It also gives you climate graphs for three places. Climate graphs are used to illustrate the type of statistics in Figure 3.2. Temperature is shown as a red line and is read off from the left axis. Rainfall is shown as blue bars and is read off from the right axis. The questions in the Question box will help you to understand these graphs.

Figure 3.3 British Isles: climate

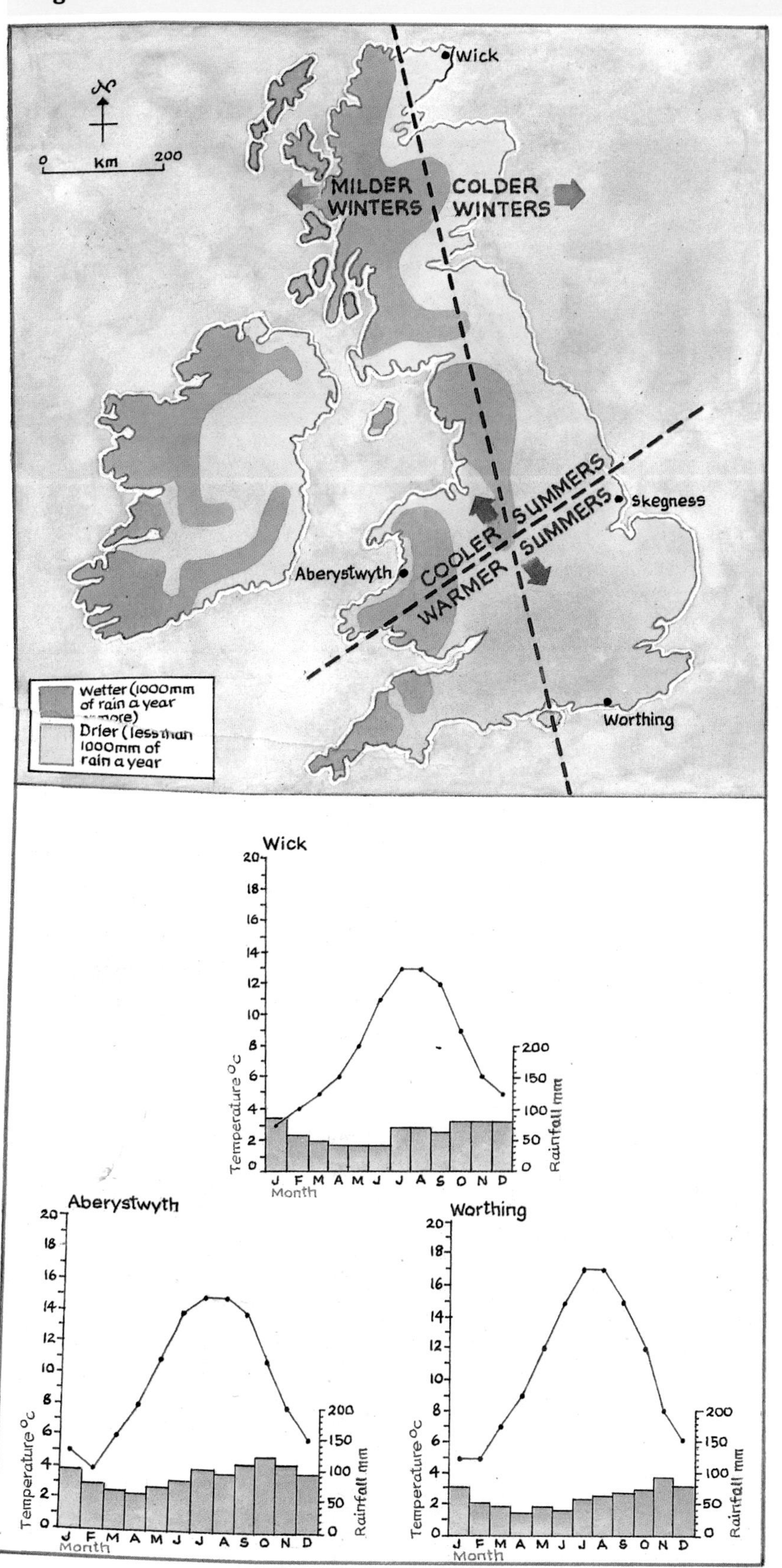

QUESTION BOX

1 Look at Figure 3.1. Which day do you think had the best weather? Why?

2 Look at Figure 3.2.

a) Which months are the warmest, the coldest, the wettest and the driest in Skegness?

b) What is the yearly **range of temperature** in Skegness?

c) What is the **total yearly rainfall** in Skegness?

3 Look at Figure 3.3.

a) Which part of the British Isles is the wettest – north, south, east or west?

b) Which part of the country is warmest in the summer?

c) Which part of the country is warmest in the winter?

A Use the statistics in Figures 3.1 and 3.2 to help you explain the difference between weather and climate.

B Use the information in Figure 3.3 to describe the pattern of climate in the British Isles.

Extension task: see worksheet 3A/B

Explaining the climate of the British Isles

Figure 3.4 Relief or 'mountain' rain

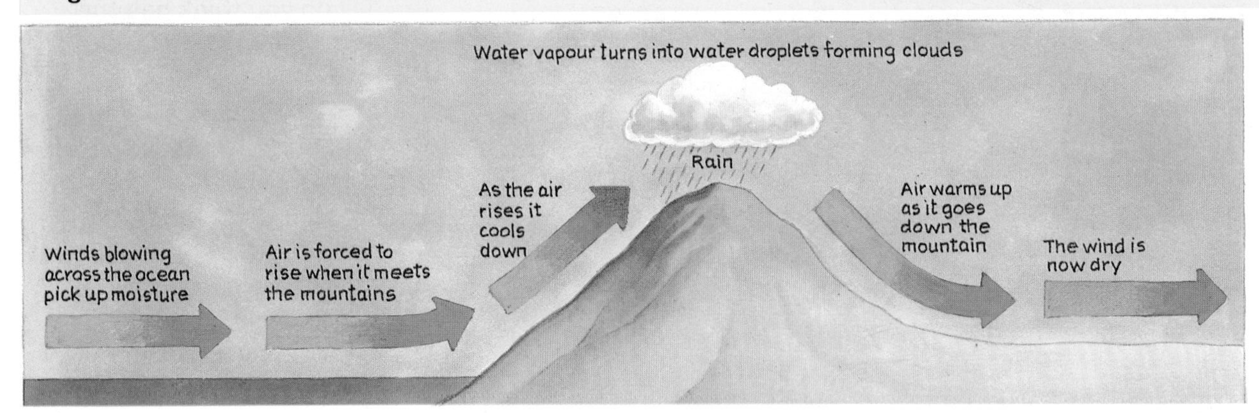

Rainfall

The west of the British Isles is wetter than the east. To explain this you have to know about the different types of rainfall and the **relief** of the land.

It rains when air is forced to rise. As it rises it cools down and **water vapour** turns into water droplets. These droplets make clouds and if they become heavy enough they fall to the earth as rain.

Relief rain (orographic or 'mountain' rain) happens when air is forced to rise over a range of mountains. Figure 3.4 shows how this happens. If you look at a relief map of the British Isles in an atlas you will see that the west is much hillier than the east. This means that the west gets much more relief rain than the east and this is one reason why the west is wetter.

Convection rain ('hot air' rain) happens when the air at the earth's surface becomes hot enough for it to rise up into the atmosphere (Figure 3.5). This type of rain is common in tropical regions but it also happens in the warmer parts of the British Isles in the summer. This helps to explain why summers in places like Skegness are as wet as the winters. Convection rain usually falls as short, heavy thunderstorms.

Figure 3.5 Convection or 'hot air' rain

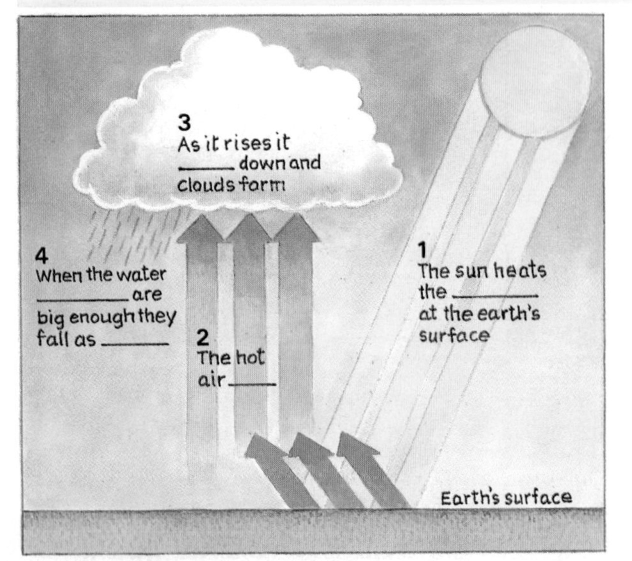

Word box

relief The height and shape of the land
water vapour Water in the air as a gas which cannot be seen
relief rain Rain caused by air rising over mountains
convection rain Rain caused by currents of hot air rising and cooling
depression rain Rain caused by warm air rising over cold air
air mass A body of air which covers a very large area and has its own special type of weather
front The place where warm and cold air meet

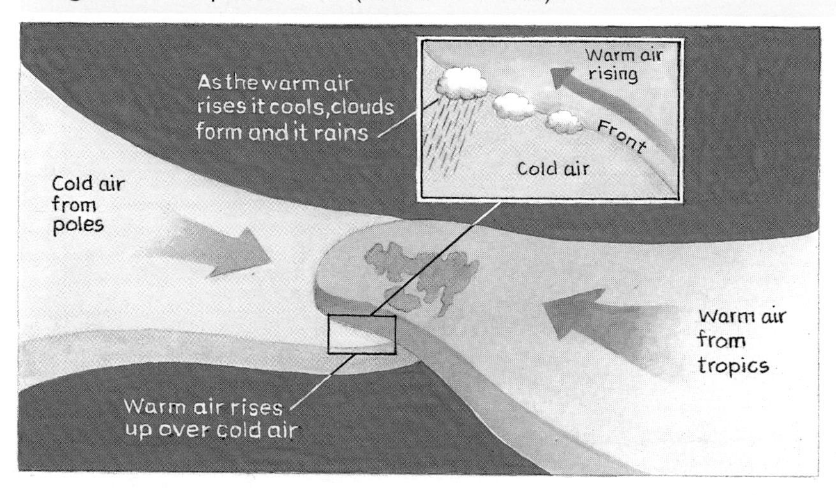

Figure 3.6 Depression rain (frontal or 'V' rain)

Depression rain (frontal or 'V' rain) happens when a warm **air mass** comes into contact with a cold air mass and rises. This situation is quite common over the British Isles because they are in a position where warm air coming up from the tropics meets cold air coming down from the poles (Figure 3.6). As the warm air rises it cools down, clouds form and it rains.

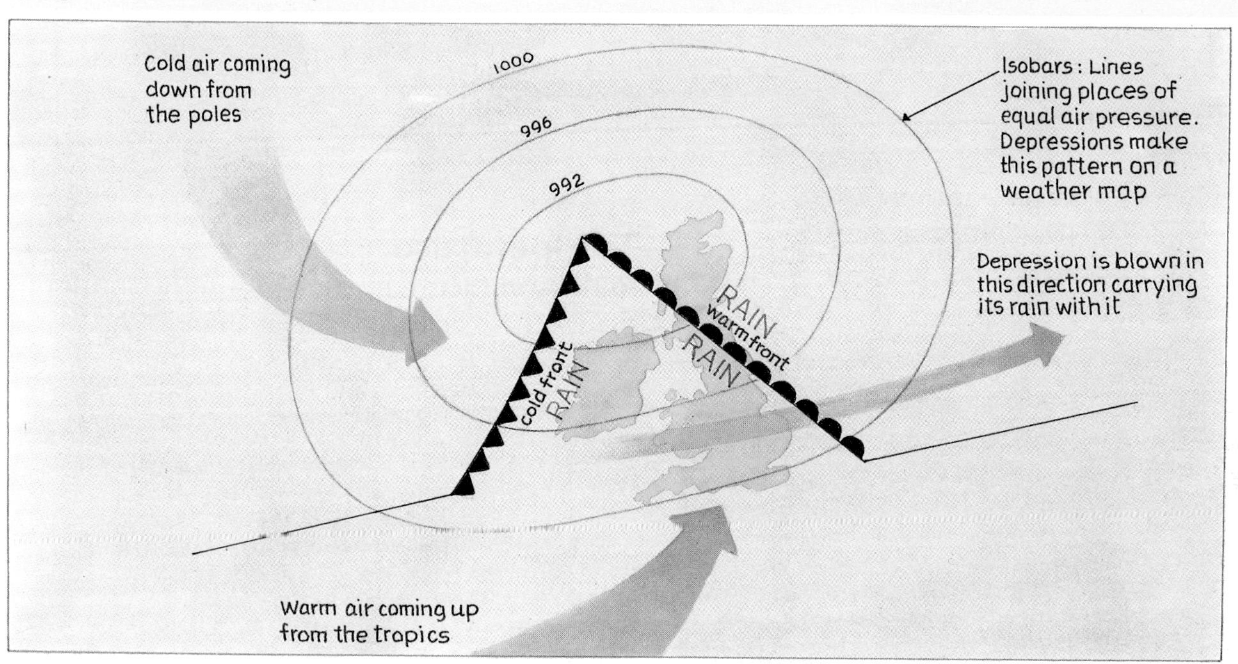

Figure 3.7 Fronts and depressions

The places where warm and cold air masses meet are called **fronts**. These fronts make a 'V' shaped pattern on a weather map so they are easy to spot (Figure 3.7).

Depressions are blown across the British Isles by westerly winds from the Atlantic. The western side of the country is open to the full force of a depression which is another reason why it is wetter than the east. However, depressions can carry rain to any part of the country.

QUESTION BOX

1 Copy and label Figure 3.5, filling in the blanks.

2 Give one reason why the west of the British Isles is wetter than the east.

A Why is Skegness more likely to get convection rainfall than Aberystwyth?

B Explain why the west of the British Isles is wetter than the east.

Extension task: see worksheet 3C/D

Figure 3.8 The apparent movement of the sun

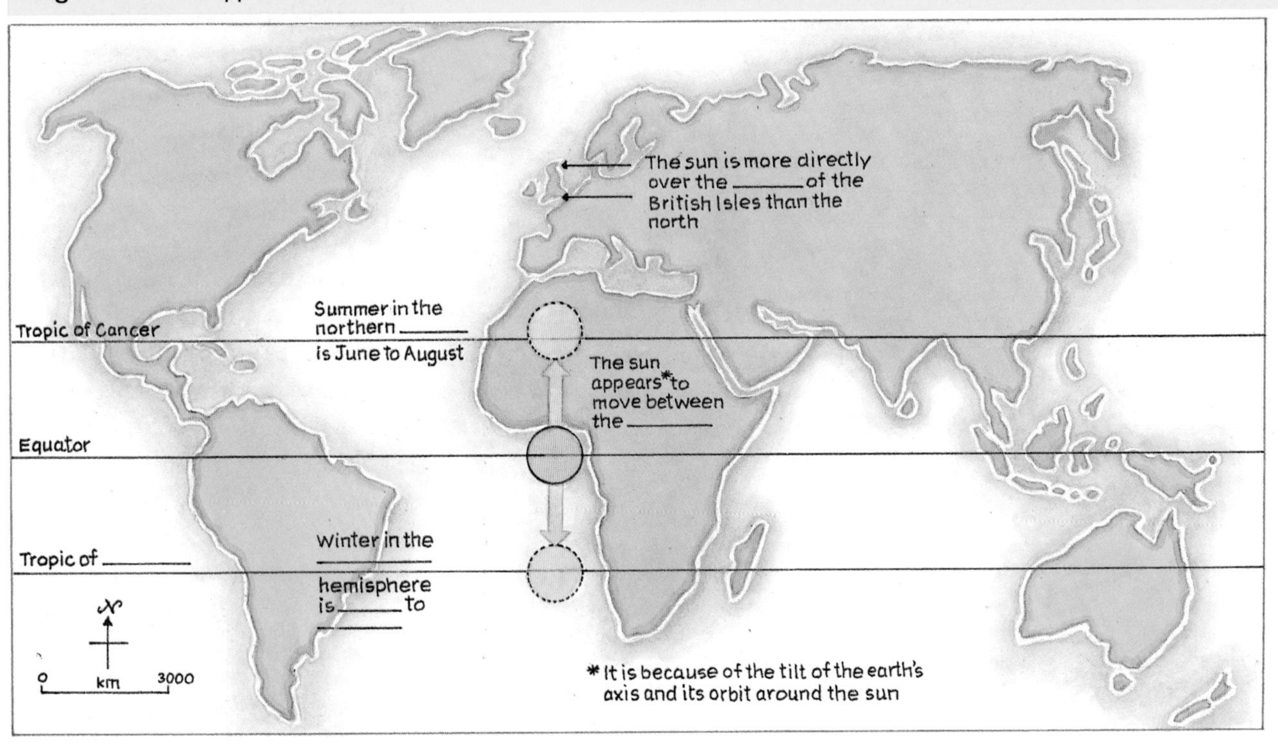

Figure 3.9 Inverewe gardens, north west Scotland

Temperature

In the summer, temperatures in the south of the British Isles are warmer than in the north. This is because the sun is more directly over the south of the country. This comes about because it is during the summer months that, on account of the tilt in the earth's axis, the sun appears to move northwards towards the tropic of cancer (Figure 3.8). As it does so, temperatures rise in the northern **hemisphere**, while it becomes winter in the southern hemisphere.

Word box

hemisphere Half of the earth

North Atlantic Drift A warm current of water which brings mild weather to the shores of north west Europe in the winter

Figure 3.10 Air masses

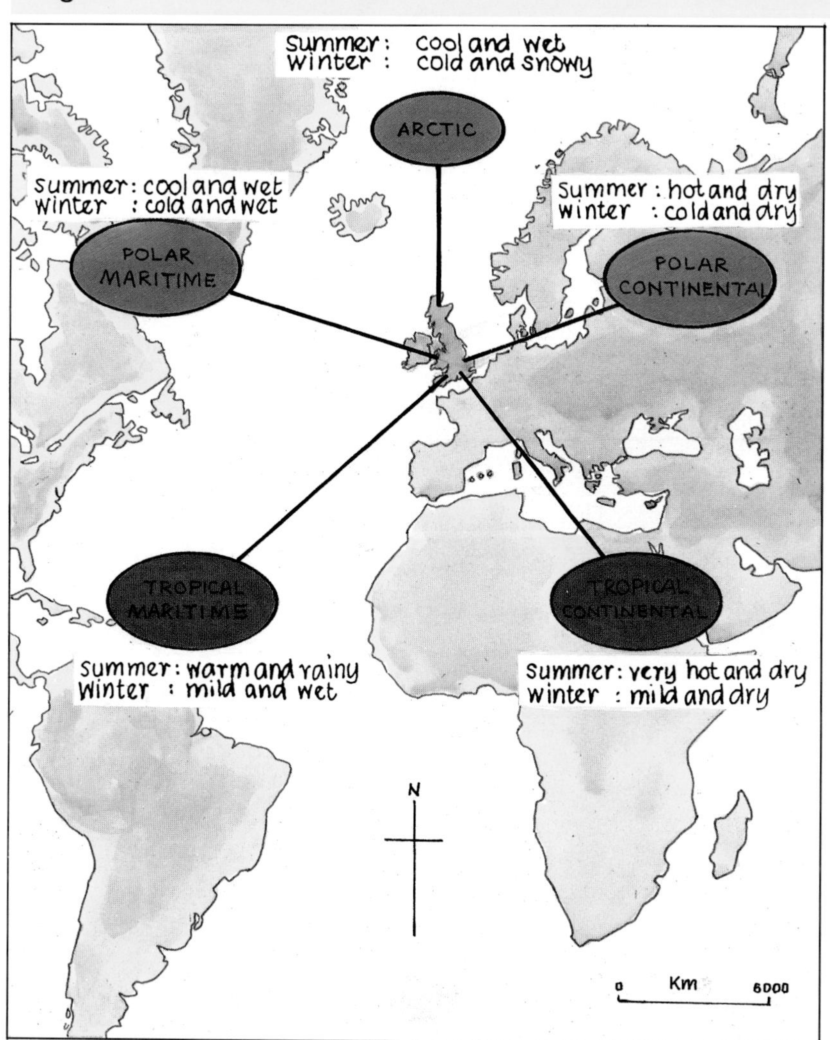

In the winter the sun is much lower in the sky, the days are shorter and temperatures are generally lower. However, a warm current of water – called the **North Atlantic Drift** – keeps the west of the country milder than the east. It travels from the Caribbean Sea, north to the Gulf of Mexico, east to the Atlantic Ocean and so to the west coast of Britain. The effects of this current really are quite remarkable: for example, sub-tropical gardens can survive in the very north west of Scotland (Figure 3.9).

Air masses

Weather in the British Isles is very changeable because winds can blow from one of five different air masses (Figure 3.10). These winds bring the type of weather you get with the particular air mass. In the summer, winds from the tropical air masses are more common and in the winter, winds from the polar air masses are more common. This is because as the sun moves between the tropics it drags the air masses with it. However, it is possible for winds from any of these air masses to blow across the British Isles at any time of the year.

QUESTION BOX

1 Mark onto a world outline the details shown on Figure 3.8. Copy and complete the labels.

2 Why is the south of the British Isles warmer than the north in the summer?

A How does the North Atlantic Drift affect the climate of the British Isles?

B Why is the British Isles affected by so many air masses? How do these air masses affect the climate of the British Isles?

Weather and tourism in the European Union (EU)

When we decide where to go on holiday we think about a lot of different things, some of which are shown in Figure 3.11. For many people what the weather is going to be like is very important. If you want to relax on a beach, it is not much fun in the rain!

The weather helps to explain the pattern of tourism in the EU. Figure 3.12 shows that most of the major coastal tourist resorts are in the warmer, drier south of the Union. The five Mediterranean countries – Portugal, Spain, France, Italy and Greece – account for nearly 60 per cent of the EU's foreign tourists (1994). This makes tourism a very important industry, earning these five countries many millions of pounds.

Figure 3.11 Choosing a holiday

Figure 3.12 Major coastal resorts in the EU

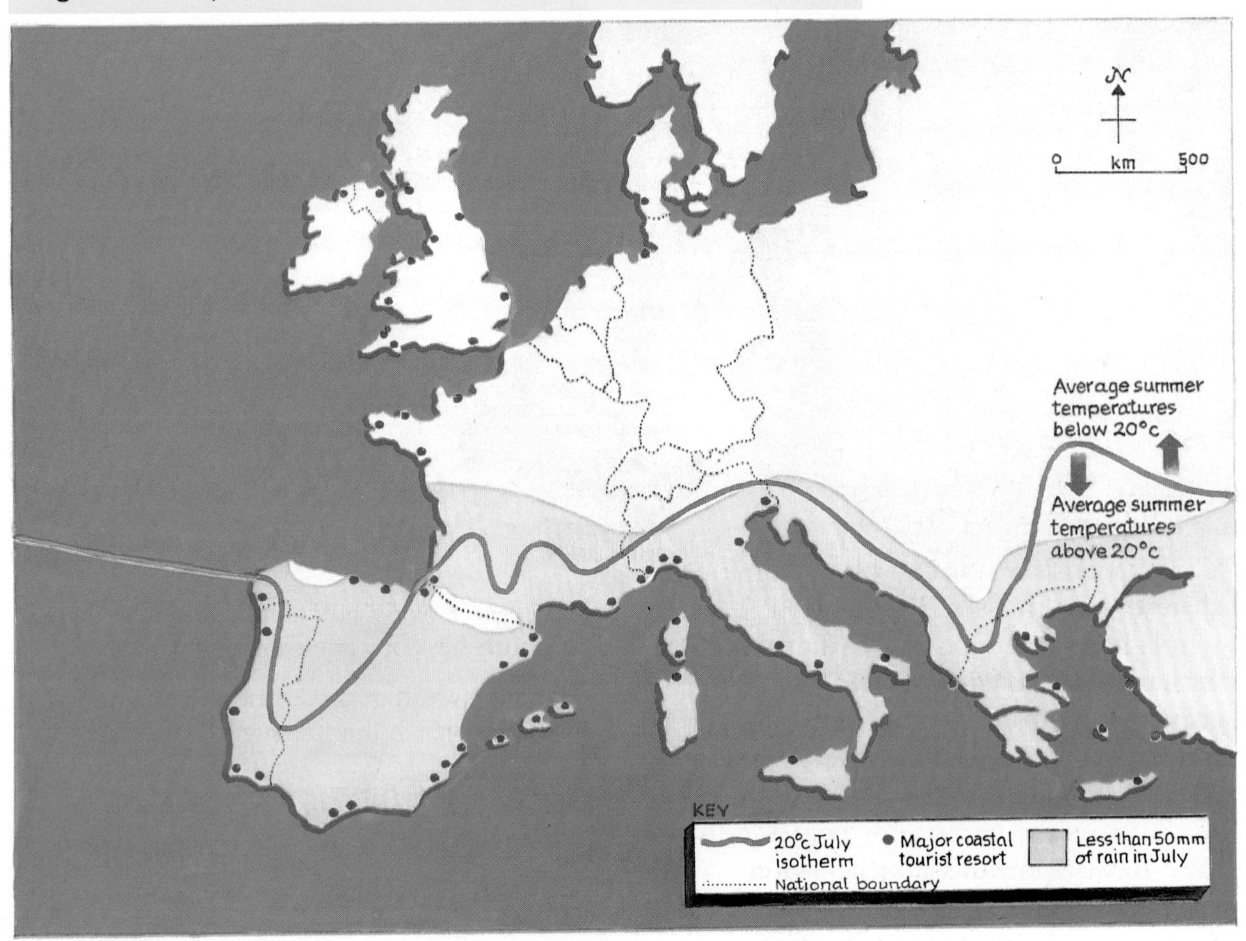

Tourism in Italy

People visit Italy for different reasons. It has famous cultural cities such as Rome and Florence; there are lakes and mountains; and it has superb beaches. However, its weather is very important in attracting foreign tourists. In the winter, snow in the Alps provides excellent skiing while in the summer its hot, sunny days have helped a number of major coastal tourist resorts to develop (Figure 3.13).

Tourism brings with it a number of advantages. Twenty million people visit each year and earn the country a profit of 10–12 billion lire (late 1990s). It also brings developments such as new roads which are of benefit to the local people as well as to the tourists.

Figure 3.14 Rimini, Italy

Tourism also brings disadvantages. Beautiful stretches of coastline have been spoilt by building miles of hotels. Litter is a problem on crowded beaches like those at Rimini on the Adriatic Riviera (Figure 3.14). **Seasonal unemployment** is a problem because, for example, most of the jobs in the coastal resorts are for the summer only.

However, the tourist industry is likely to become even more important to the Italian economy in the next few years. The government is using tourism to spread **economic development** into the poorer south of the country by building hotels and roads in areas with fine beaches and hot, sunny summers (see Figure 3.13).

Word box

seasonal unemployment This is when people are out of work because their job is for one season of the year only, e.g. a ski instructor

economic development This is when a country or region sets up new industries and becomes wealthier

isotherm A line joining places of equal temperature

Figure 3.13 Tourism in Italy

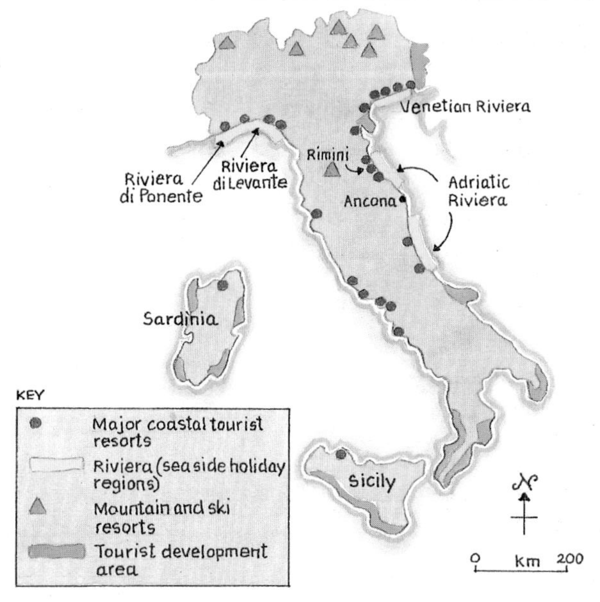

QUESTION BOX

1. What type of weather is important to tourism in a) the Italian Alps and b) coastal resorts such as Rimini?
2. Give one advantage and one disadvantage of tourism in Italy.

A How is the weather important to tourism in Italy?

B Do you think that building hotels is a good way to spread economic development in Italy? Explain your answer.

Hurricane!

Hurricanes are an example of the weather as a **natural hazard**. They are circular in shape and can cover an area up to 800 km in diameter. The centre – or eye – of a hurricane is usually about 30 km in diameter and gives an area of clear skies and calm winds. However, winds spiral in towards the centre at speeds of up to 300 km per hour (km/h) and they bring with them rain, thunder and lightning. Hurricanes always begin over warm tropical seas and move slowly towards the land. The sea is the source of their energy so they fade away as they move inland.

In September 1998 Hurricane Georges swept through the Caribbean and the Gulf of Mexico (Figure 3.15). It hit Puerto Rico on 21 September, killing 3 people and causing $2 billion worth of damage. Wind speeds were even greater when it crossed the Dominican Republic, leaving 210 dead and destroying the country's plantations (Figure 3.16). Eighty seven were killed in Haiti and 4 were killed in Cuba.

On 25 September it crossed the Florida Keys. It destroyed house-boats and tore off roofs but no-one was killed, largely because 1.4 million people had been advised to leave their homes and drive north to get away from the storm.

Hurricane Georges carried on across the Gulf of Mexico and it seemed to be heading towards New Orleans. An even worse disaster was feared because New Orleans is built below sea level and is only protected by artificial **levées**. One and a half million people were evacuated but it eventually came ashore to the east of New Orleans at Biloxi on 28 September. It caused a great deal of damage although only one person died. The storm finally died out as it moved inland.

Figure 3.15 Tracking Hurricane Georges

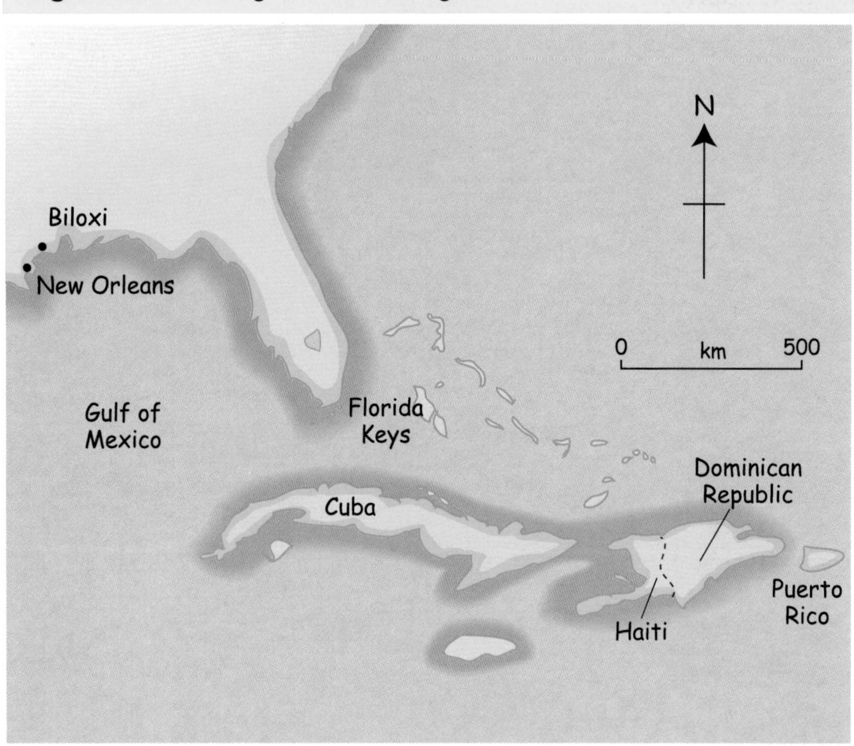

Figure 3.16 Hurricane Georges in the Dominican Republic

Word box

hurricanes Violent tropical storms, also known as cyclones (Indian Ocean) and typhoons (Pacific Ocean)

natural hazard A violent natural event which threatens people

levée Originally a natural river bank formed by deposits of sediment following flooding, many have been artificially raised to form river defences

QUESTION BOX

1 Draw and label a map to show the path taken by Hurricane Georges and the damage and loss of life it caused.

A Why do you think loss of life was so much greater in the islands of the Caribbean than in the USA?

SUMMARY QUESTIONS

1 Draw climate graphs for Cortina d'Ampezzo, which is in the Italian Alps, and Sorrento, which is on the coast near Naples (Figure 3.17).

2 Describe a) the temperature and b) the rainfall at each of these two places.

3 Why do you think it is so much colder in Cortina d'Ampezzo than in Sorrento in the winter?

4 Both places are tourist resorts. What sort of holiday do you think they offer? Suggest reasons for your answer.

A Draw climate graphs for Cortina d'Ampezzo, which is in the Italian Alps, and Sorrento, which is on the coast near Naples (Figure 3.17).

B Describe the climate of each of the two places.

C What do you think is the main reason for their climates being different? Explain your answer.

D Both places are tourist resorts. What sort of holiday do you think they offer, and why? How will the economies of these places be affected by their involvement with tourism?

Figure 3.17 Choosing a holiday

a) Cortina d'Ampezzo (1800 m above sea level)

Month	J	F	M	A	M	J	J	A	S	O	N	D
Temp °C	–2	–1	2	5	9	14	16	16	13	8	3	–2
Rain mm	50	45	80	140	130	150	150	115	115	120	115	60

b) Sorrento (30 m above sea level)

Month	J	F	M	A	M	J	J	A	S	O	N	D
Temp °C	9	9	11	14	17	21	24	24	21	17	13	10
Rain mm	120	90	65	60	50	20	15	25	90	120	130	135

unit 4

THE GREEN PLANET

Key questions

- What is an ecosystem?
- What are the characteristics of the savanna grassland?
- What types of farming are found in the African savanna? How does farming affect the savanna ecosystem?
- How does tourism affect the savanna ecosystem?
- What is desertification and why is it a problem in the savanna grassland?

Ecosystems

Figure 4.1 a) Church Green pond, Ramsey b) Wistow Wood, Cambridgeshire c) Earth from space

An ecosystem consists of the plants and animals of an area, and all the things which make up their surroundings, like soil, water and air. It can be a small area such as a pond; a medium-sized area such as a wood; or a large area such as the earth itself (Figure 4.1).

Word box

food chain Plants being eaten by animals which are, in turn, eaten by other animals etc
herbivores Animals which eat only plants
carnivores Animals which eat only meat
energy flow Energy moving through an ecosystem
decomposers The bacteria and insects which break down dead plants and animals
nutrient cycling 'Goodness' moving through an ecosystem

Ecosystems are made up of living parts, such as plants, and non-living parts, such as soil. These parts are linked together; for example, plants take nutrients (goodness) from the soil.

An example of an ecosystem is the African savanna (Figure 4.2). Grasses are the main type of vegetation. They can grow up to 5 m tall in the rainy season but die back in the dry season. Trees, which are dotted about the landscape, have adapted (learnt to cope with) the dry season in different ways; for example, the baobab tree stores water in its trunk.

Many types of animal live in the African savanna and they are linked together as part of a **food chain**. **Herbivores**, such as zebras, graze on the grasses. **Carnivores**, such as lions, eat the herbivores. Other carnivores, such as vultures, scavenge on half eaten carcasses.

The grasses and other plants 'trap' energy from the sun and store it in their leaves. When the grasses are eaten energy is moved through the ecosystem along the food chain. This is known as **energy flow** and is very important in the working of an ecosystem.

The heavy rain in the savanna's wet season washes many of the nutrients out of the soil. However, insects, such as termites, drag dead plant and animal matter underground where **decomposers** break it down until it is small enough to be taken up by plants. This movement of 'goodness' through an ecosystem is known as **nutrient cycling**.

QUESTION BOX

1 Copy Figure 4.2, but only the labels, not the pictures. Put a green box around the labels which show the living parts of the ecosystem. Put a blue box around the labels which show the non-living parts. Put a red box around the labels which say how the ecosystem works.

A Describe and explain the main components * and processes ** of the savanna ecosystem.

* what it is made up of

** how it works

Extension task: see worksheet 4A/B

Figure 4.2 The African savanna

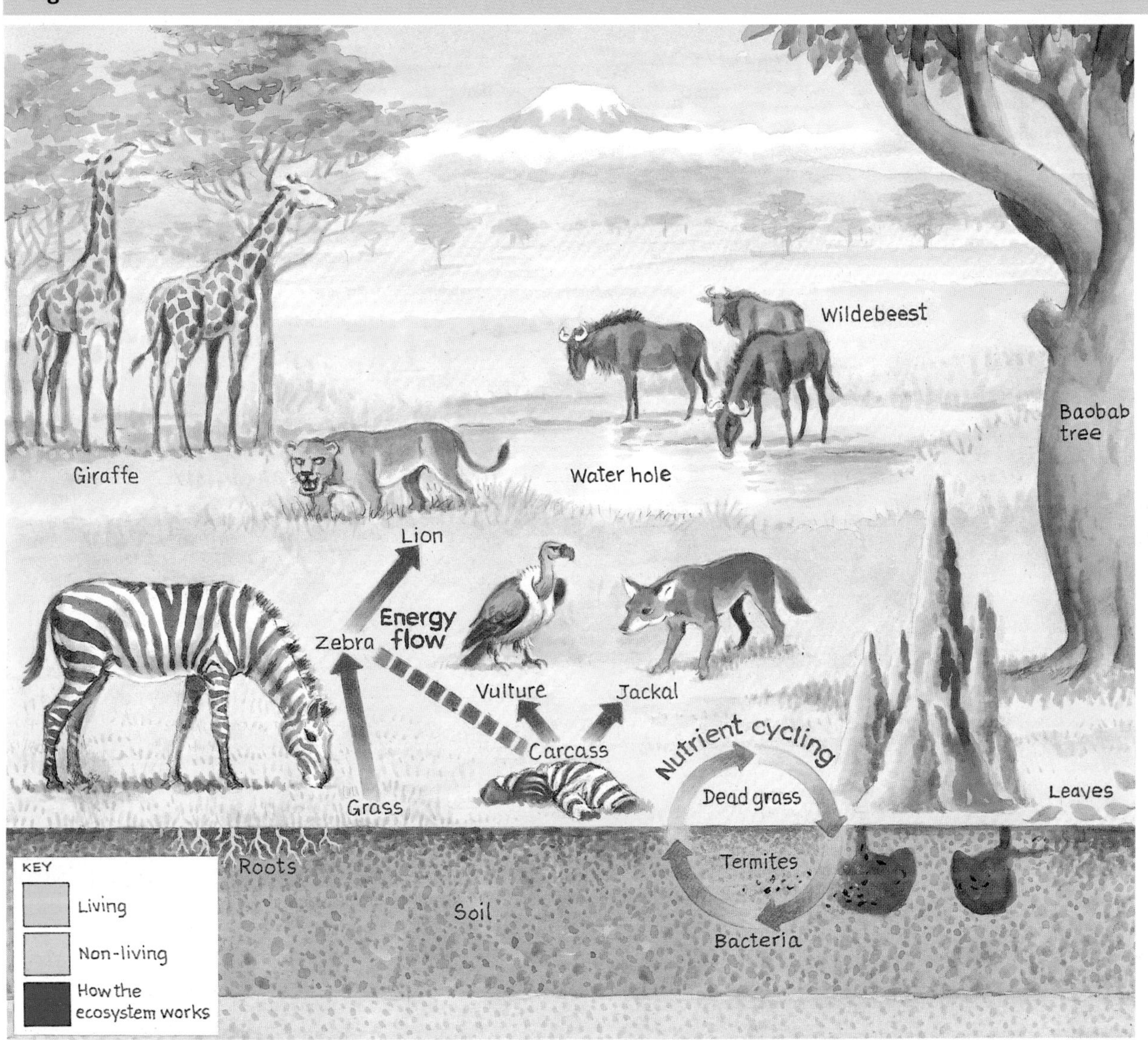

The savanna ecosystem

The savanna ecosystem covers large parts of the tropics, not only in Africa, but also in South America and Australia (Figure 4.3).

Savanna regions have the same type of climate. There are two seasons: one is hot and dry and the other is hot and rainy. Kano, in Nigeria, is a good example of this type of climate (Figure 4.6); in the winter there is hardly any rain at all; whereas in the summer **convection rain storms** give high monthly totals.

These regions also have the same type of soil. It is a red/brown colour because it has been stained by iron. This happens because high temperatures and heavy rain in the wet season cause rapid **chemical weathering**. The heavy rain also washes away a lot of the goodness in the soil, a process known as **leaching**. Fortunately, these nutrients are replaced by the work of the decomposers.

Grasses are the main type of vegetation because they are so good at coping with the difference between the two seasons (Figures 4.4 and 4.5). In the dry season the grasses die back and use their long roots to reach down for any water left in the soil. In the wet season they grow, flower and spread their seeds quickly.

Figure 4.3 World distribution of the savanna ecosystem

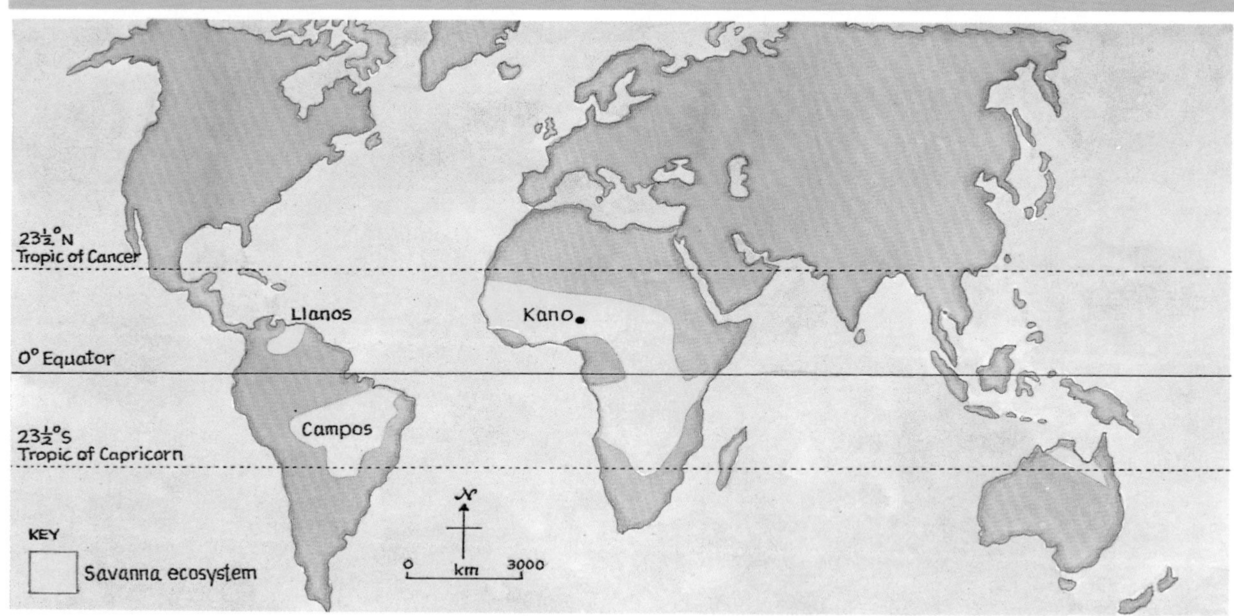

Figure 4.4 Dry savanna in south east Bolivia

Figure 4.5 Wet savanna in south east Bolivia

Trees find the savanna conditions more difficult to cope with. Some, like the baobab, store water in their trunks. Others, like the acacia, have thorns to protect themselves from animals in search of moist leaves. Large numbers of trees are only found in the wetter parts of the savanna.

Figure 4.6 Climate graph for Kano, Nigeria

Animals also have to cope with the extreme climate. In the dry season they may have to roam for many kilometres to find water. In the wet season, grazing is plentiful and they make the most of it by putting on as much weight as possible.

Fires, started naturally by lightning, are a common feature of the savanna; on average, they happen once every four years. Unless a fire gets out of control, what at first may seem like a disaster is, in fact, very important to the ecosystem. It gets rid of dead vegetation and allows new shoots to grow through. Many plants have adapted so that their seeds will only open if they are heated. Ash from the burnt vegetation is also important as a source of fertiliser.

Word box

convection rain storms Rain caused by currents of hot air rising and cooling

chemical weathering The disintegration of rock because of a chemical reaction, e.g. the oxidisation (rusting) of iron minerals by rain water

leaching Nutrients and minerals being washed out of the soil by rain

QUESTION BOX

1 Look at Figure 4.3. What names are used for the savanna ecosystem in South America?

2 Look at Figure 4.6. What is a) the highest and b) the lowest temperature in Kano? What is the total annual rainfall?

3 Use Figures 4.4 and 4.5 to help you describe the differences between the savanna's two main seasons.

A Use Figures 4.4, 4.5 and 4.6 to describe the similarities and differences between the savanna's two main seasons.

B Explain how plants and animals have adapted to the savanna climate.

C Why is fire important to the savanna ecosystem?

People and the savanna ecosystem

Farming in the African savanna

The main problem with which farmers in the African savanna have to cope is the climate. The rains at the end of the dry season are unreliable and if they are late both crops and animals suffer. Also, the heavy rains in the wet season cause leaching (see page 36) and wash away the soil if it is left unprotected. The farming systems found in the savanna manage these conditions in different ways.

Figure 4.7 Shifting cultivation

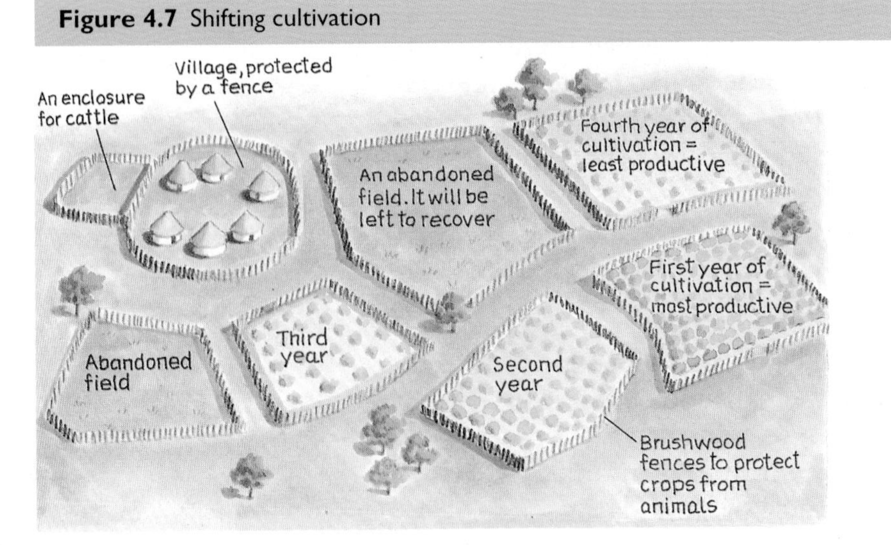

Shifting cultivation, a type of **subsistence farming**, is the traditional way of growing crops. The shifting cultivator uses a field for three or four years, until it begins to lose its fertility, and then leaves it for a new one (Figure 4.7). One of the big advantages of farming in this way is that it gives the soil a chance to recover but a disadvantage is that it takes up a large area of land.

The Hausa in northern Nigeria have farmed in this way for hundreds of years. In February and March, before the rains come, the fields are cleared by burning. In April, the fields are ploughed and as soon as the rainy season begins, in May or June, the crops are planted. Yams, millet, maize, groundnuts and beans are grown. Harvesting is in October and November, when the rains have stopped. If a field is beginning to lose its fertility, a new area of land will be cleared in the following spring.

Nomadic herding, another type of subsistence farming, is the traditional way of keeping cattle. Nomadic herders, such as the Masai in East Africa (Figure 4.8), migrate with their animals in search of grazing and water. In the dry season they may have to travel many

Figure 4.8 Masai herders, Kenya

Figure 4.9 Wheat farm in Tanzania, Africa

hundreds of kilometres every month. They drink the milk and blood of their cattle but they do not kill them for meat. They hunt wild animals and gather fruit, nuts and berries. As with shifting cultivation, this type of farming looks after the land but needs a very large area.

Other types of **commercial farming** have been developed in the African savanna in recent years. For example, the Canadian Government has helped to set up a wheat growing scheme in Tanzania on the Hanang Plains (Figures 4.9 and 4.10). The aim is to make the country self-sufficient in wheat and, to date, over 30 000 ha have been planted. It will mean that Tanzania will no longer have to pay for **importing** wheat. However, machinery, such as combine harvesters, is expensive and has to be imported, and also the scheme is in an area where Masai herders used to graze their animals.

Figure 4.10 Location map: the Hanang Plains, Tanzania

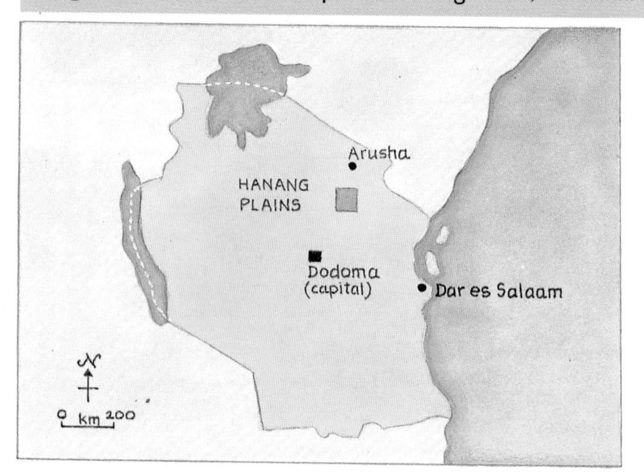

Figure 4.11 Kenyan tea plantation

Word box

shifting cultivation Abandoning a field after three or four years, to let it recover, and moving on to farm new land

subsistence farming Producing food for the family to eat. If there is any left over it is sold to buy luxuries such as soap

nomadic herding Where people migrate with their animals in search of grazing and water

colonies Lands occupied and ruled by another country

export A product sold to another country

commercial farming Producing food mainly to sell for money

importing Buying goods from another country

In the nineteenth century many countries in Africa were taken over as **colonies** by the UK, France and Germany. Plantations – large farms which grow just one crop, usually for **export** – were set up. Plantation crops include tea (Figure 4.11), coffee, tobacco and cotton and they are still very important; for example, Kenya's main exports are coffee and tea and Tanzania's are coffee and cotton. Plantations place a strain on the land because the same fields are used all the time. However, they are an important source of income.

QUESTION BOX

1 Draw and label a diagram to show the main events in the shifting cultivator's year.

2 Give one advantage and one disadvantage of each type of farming mentioned in this section.

A Explain how a) shifting cultivation and b) nomadic herding are adapted to the savanna ecosystem.

B How and why is farming changing in the African savanna? Who might benefit and who might lose from these changes?

Extension task: see worksheet 4C/D

Tourism in Kenya

Kenya has spectacular scenery and superb beaches but most tourists visit it to see the wildlife of the savanna grasslands (Figure 4.12). There are 40 game parks and reserves, covering eight per cent of the entire country, the largest of which are shown in Figure 4.13.

Figure 4.12 Giraffe in Kenyan grasslands

Figure 4.13 National parks and game reserves in Kenya

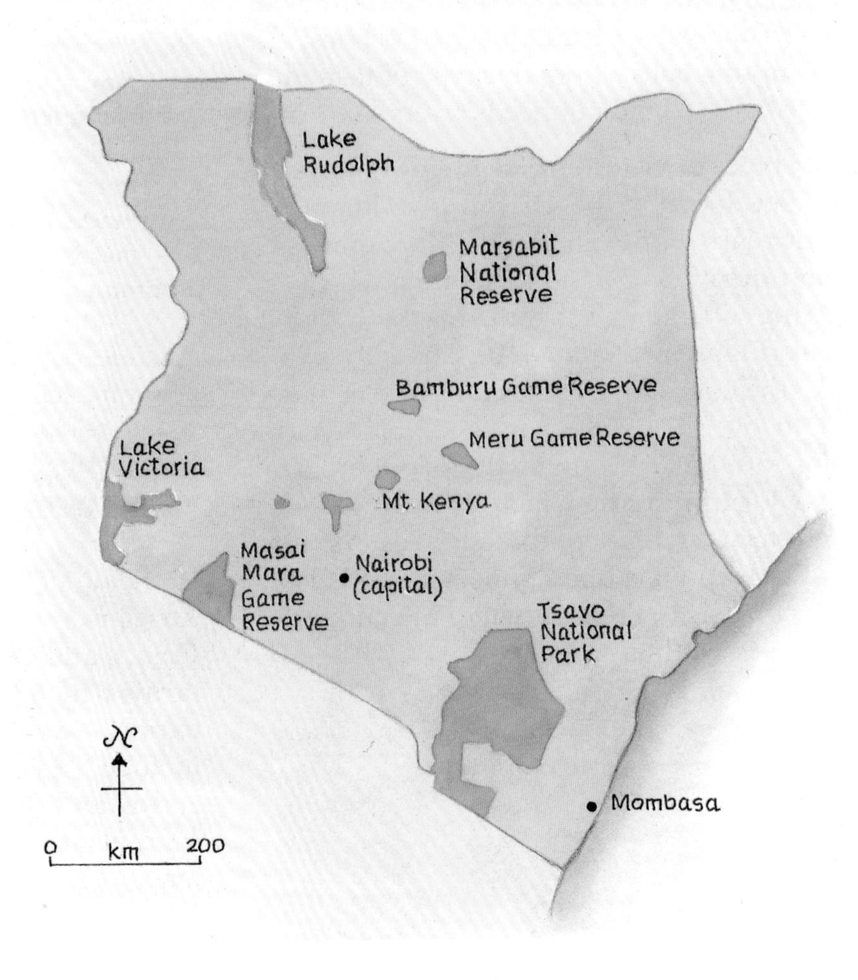

Tourism has brought a number of benefits. It earns US$375 million from its 800 000 visitors a year. It has spread development away from the main cities of Nairobi and Mombasa. It employs 350 000 people, and training colleges have been set up so that Kenyans can learn the skills needed by the tourist industry, such as hotel management and catering.

However, it has also brought a number of problems. There are worries about the animals. Rhinos have been scared out of the Masai Mara by the hot air balloons used by tourists to observe the wildlife. Baboons have learnt to scavenge food from hotel dustbins and, as a result, they have become overweight and unhealthy. Tourist buses often get too close to the animals and this has had a bad effect, in particular on lions and cheetahs who are hunting and breeding less.

Tourism has also caused problems for many of the local people. **Nomadic herders**, such as the Masai, are no longer allowed to graze their cattle in the game reserves. Instead, they have been forced to settle in villages around the edge of the reserves. **Overgrazing** has become a problem because they have not been given enough land for their cattle and this, in turn, has led to **soil erosion**

Word box

nomadic herders People who migrate with their animals in search of grazing and water

overgrazing Animals eating all the vegetation so that there is none left

soil erosion The soil being worn away, e.g. by rain or wind

(Figure 4.14). Also, the fences built to protect their land have made it difficult for wild animals to migrate in the dry season in search of water.

There have been attempts to deal with some of these problems. The Kenya Wildlife Service (KWS) has been improved a great deal since 1989; for example, the number of people working for it has risen from a few hundred to over 4000. It has stopped a great deal of poaching and it now controls where tourists are allowed to go.

Local people have become more involved in running the reserves. This gives them a source of income to make up for the loss of their traditional grazing lands. It is also a good idea because they know a great deal about the savanna ecosystem and how it should be looked after.

The government has also tried to help the villages on the edge of the reserves by spending more of the money made from tourism on community projects, such as making souvenirs (Figure 4.15), and on better roads and services.

QUESTION BOX

1 Why do tourists visit Kenya?

2 Give one advantage and one disadvantage of tourism in Kenya.

A How has the development of tourism in some parts of Kenya led to overgrazing and soil erosion?

B Why is it a good idea to involve local people in running Kenya's game reserves?

Figure 4.15 Masai craft shop, Tsavo National Park

Figure 4.14 Overgrazing and soil erosion

Healthy grazing (a)
Leaves
Roots
Soil
Rock

Overgrazing removes all the vegetation (b)
No leaves to protect soil from rain
No roots to bind soil together

Land degradation (d)
Soil is completely eroded. Grazing is destroyed

Soil erosion (c)
Soil is washed away

Desertification

Desertification is when a desert grows and takes over farmland. It has become a big problem since the 1970s in the zone between the hot deserts and the savanna grasslands. The worst affected area has been the Sahel, a region to the south of the Sahara desert (Figure 4.16).

No one is quite sure why desertification is happening. Some think that the climate is becoming drier. It is true that there have been long periods of **drought** in the savanna in recent years; for example, there were droughts in the Sahel between 1964–75 and 1980–4, and in South Africa between 1990–2. However, rainfall in the savanna has always been unreliable with only a 25 per cent chance of a 'good' year.

Others blame population increase; for example, the population of the Sahel has grown from 19 million in 1961 to 40 million in 1989. The need to feed more people has led to overgrazing while providing fuel wood for more people has led to **deforestation**.

In turn, overgrazing and deforestation have caused soil erosion, (see pages 40–41), which has led to desertification.

Settled agriculture (always using the same fields) has become widespread, mainly in an attempt to feed the growing population, and this has also been blamed for desertification. The problem with this type of farming, compared with shifting cultivation, is that the fields do not get a chance to recover. As a result, the soil becomes exhausted and 'crumbly' which means that it is easily eroded.

It is also important to remember that most of the countries affected by desertification are very poor. This makes it extremely difficult for them to do deal with the problem, whereas a rich country like the USA can more easily afford irrigation schemes to supply water to dry land.

Desertification has led to **famine** and the deaths of millions of people. Things have been made worse in a number of countries by wars which have stopped **foreign aid** from getting through and which have forced thousands of people to leave their homes as **refugees**.

For example, a famine in the savanna of north east Kenya in 1992 was made worse by the arrival of 350 000 refugees from Somalia.

Figure 4.16 Location map: the Sahel

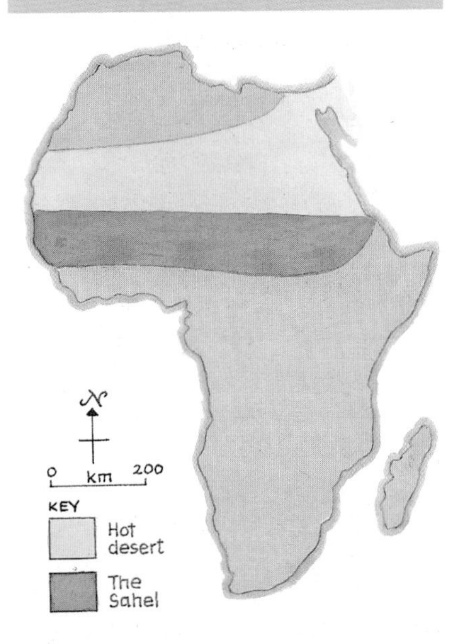

Figure 4.17 'Magic stones' – so-called because the rain they divert dramatically increases crop yields

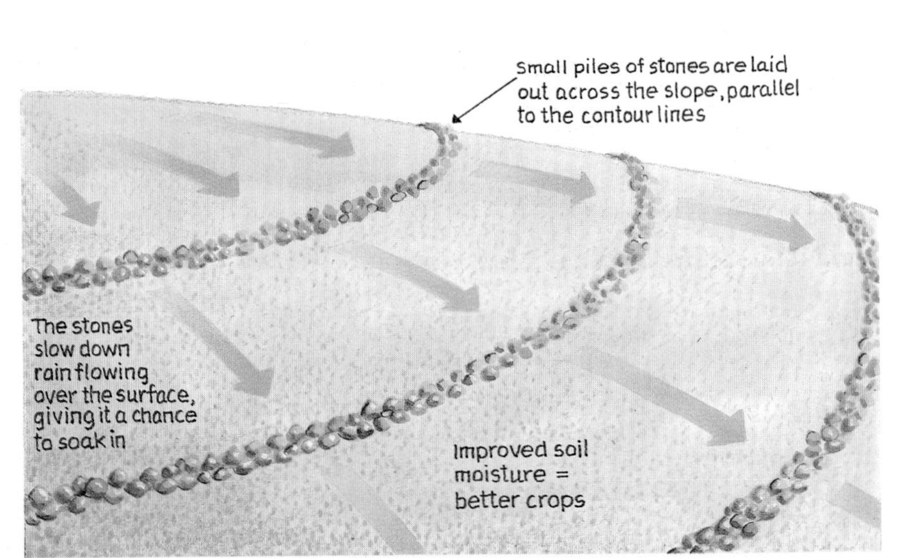

Some things can be done to hold back desertification. Aid money has been used to dig new wells and build wind pumps to raise water for **irrigation**. Small dams have been built to store water in the rainy season. It has also been discovered that if stones are lined up across a slope, they slow down the rain which would normally run straight off the hard tropical surface, and the result is a 50 per cent increase in crop yield (Figure 4.17).

Word box

drought A long spell with no rain
deforestation Completely clearing an area of trees
famine So little food that death from starvation is a possibility
foreign aid Help given by one country to another
refugee A person who has been forced to leave their home country because they are in danger
irrigation Adding water to land that is normally too dry to grow certain crops

QUESTION BOX

1 What is desertification?
2 Give one physical reason and one human reason for desertification.
3 How do 'magic stones' control desertification?

A Copy and complete Figure 4.18 which shows the possible causes of desertification.
B Describe the different ways in which the problem of desertification can be managed. Which of these methods is most likely to be successful in the African savanna, and why?

Figure 4.18 The possible causes of desertification

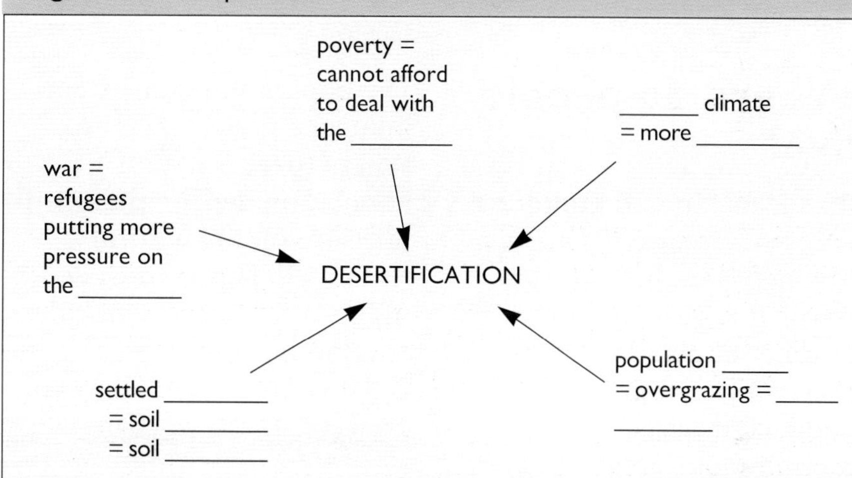

SUMMARY QUESTIONS

1 Write a report about the tropical rainforest. Include:

- a world map to show its distribution
- a climate graph
- notes about its soil
- a sketch or diagram showing the living and non-living parts of the rainforest ecosystem
- a list of human activities and any problems they cause.

Say how you think the rainforest should be used in the future. Find out your information from atlases, reference books, CD-ROMs and the Internet.

A Choose one of the world's major ecosystems, other than the savanna. Write down the questions you think you should try to answer about your ecosystem, in the order in which you think you should answer them. Get this checked before you start your research. Once your list has been approved use atlases, reference books, CD-ROMs and the Internet to research your information. Write up your information in the form of a magazine article, or as a computer/web page presentation.

unit 5

PEOPLE EVERYWHERE

Key questions

- Why is there an uneven spread of people?
- Why do hostile regions have a sparse population?
- Which regions have the highest population growth?
- How does a population structure change over time?
- What causes rural-urban drift in an LEDC?
- Why do people emigrate?
- Why do MEDCs and LEDCs have different population problems?

Where do people live?

People are spread unevenly across the planet. There is a link between the type of **natural region** and the number of people who live there. Difficult areas like hot and cold deserts, mountainous areas or rainforests have a low **population density** (Figures 5.1 and 5.2). Farming areas have a higher population. The most densely populated regions are large urban and industrial areas.

Humans are restricted in where they can live, although, technological developments mean that more people can live in difficult areas. In general, however, people live where it is neither too hot nor cold; where there is reliable rainfall, the land can be farmed, it is accessible, safe and not too high and so there is sufficient oxygen.

The population density can influence how people live. Where there is a high population density there is pressure on the natural resources. This is most acute in the cities of economically developing countries. There is a link between the level of development and how fast the population grows. **Less Economically Developed Countries** (LEDCs) have a higher population growth than **More Economically Developed Countries** (MEDCs).

Figure 5.1 World population density

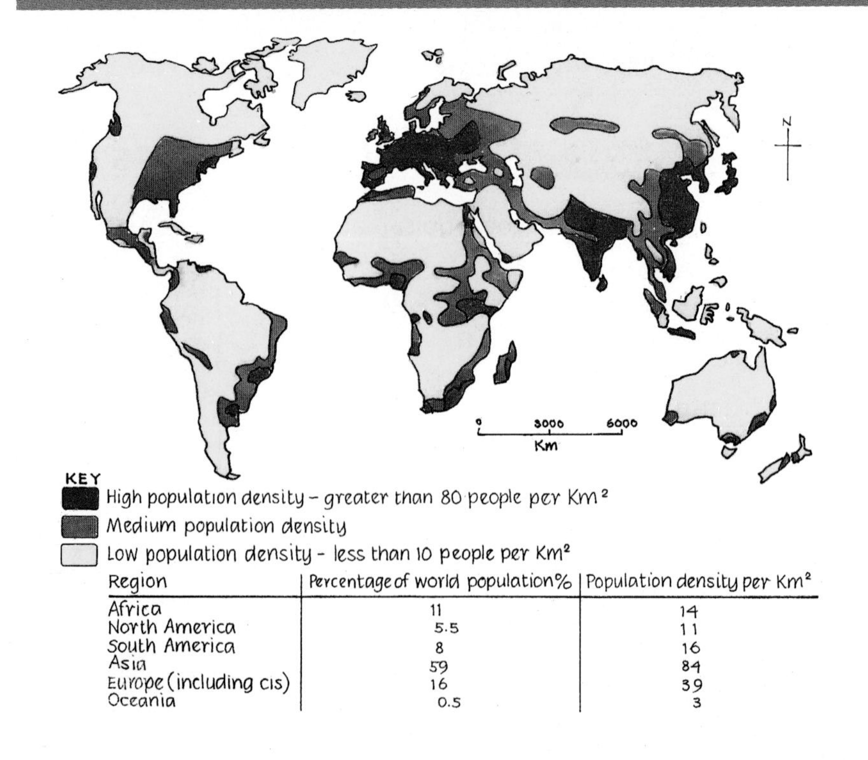

Region	Percentage of world population%	Population density per Km²
Africa	11	14
North America	5.5	11
South America	8	16
Asia	59	84
Europe (including CIS)	16	39
Oceania	0.5	3

Figure 5.2 World natural regions

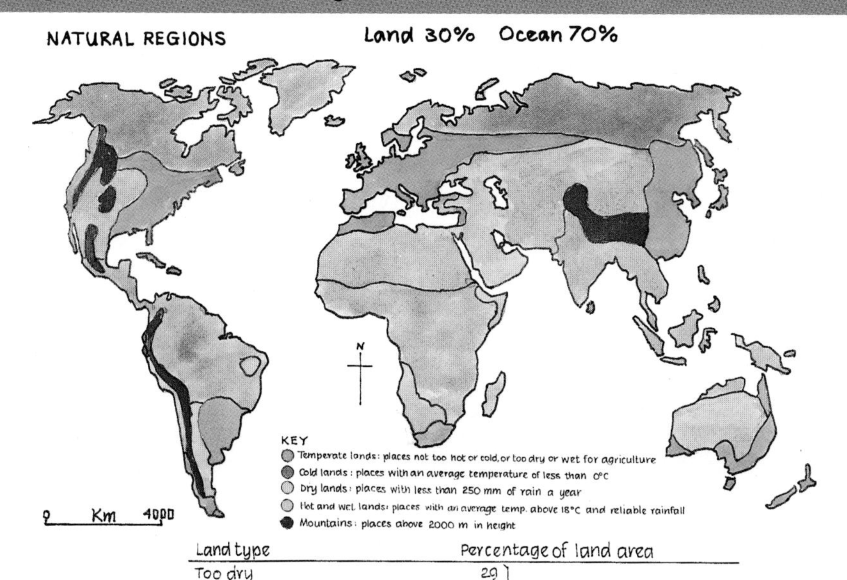

Land type	Percentage of land area	
Too dry	29	66% unsuitable for farming
Too cold	17	
Too mountainous	11	
Soils too poor for farming	9	
Land which could be used for farming	23	34% suitable for farming
Land actually used for farming	11	

QUESTION BOX

1 **a)** Trace a copy of Figure 5.1.

b) Place your copy over Figure 5.2.

c) Describe the link between the natural region and population density.

2 Draw a proportional bar chart and write a sentence to describe.

a) the land–sea percentage.

b) the percentage of land types.

3 State **three** reasons why few people live in the areas shown in Figure 5.3 and 5.4.

A Copy and complete Figure 5.5 by using Figure 5.1. Describe the distribution and pattern of world population growth.

B **a)** choose a hostile natural region.

b) on a world outline shade where this natural region is found.

c) explain why people find this a difficult region to live in.

Word box

hostile region A region which is difficult to live in, for example a hot desert, cold desert, mountain area or tropical rainforest

natural region A distinctive area with a particular climate and natural vegetation

population density The number of people per km^2

Less Economically Developed Country (LEDC) A poor country which is attempting to improve and develop its farming, industry and services in order for its population to have a higher standard of living

More Economically Developed Country (MEDC) A country which has a high Gross Domestic Product (GDP) per person, a well-developed infrastructure and a high standard of living

Figure 5.3 Himalayas, India

Figure 5.4 Kalahari Desert, South Africa

Figure 5.5 Distribution of population growth

Rank order	Region	% of world population	Population density (km^2)
1			
2			
3			
4			
5			
6			

How does a population change?

Population increases when the **birth rate** is greater than the **death rate**, and when more people move into an area than move out.

Transition theory

Population growth can be described as a series of stages. The stages can be linked to the level of a country's economic development. As a country becomes richer it is able to provide a higher **standard of living** and health care. The death and birth rates fall and population growth slows (Figure 5.6). Small family sizes are favoured as more children are likely to survive childhood. With better health care, more people live longer and many diseases which were once fatal can be cured.

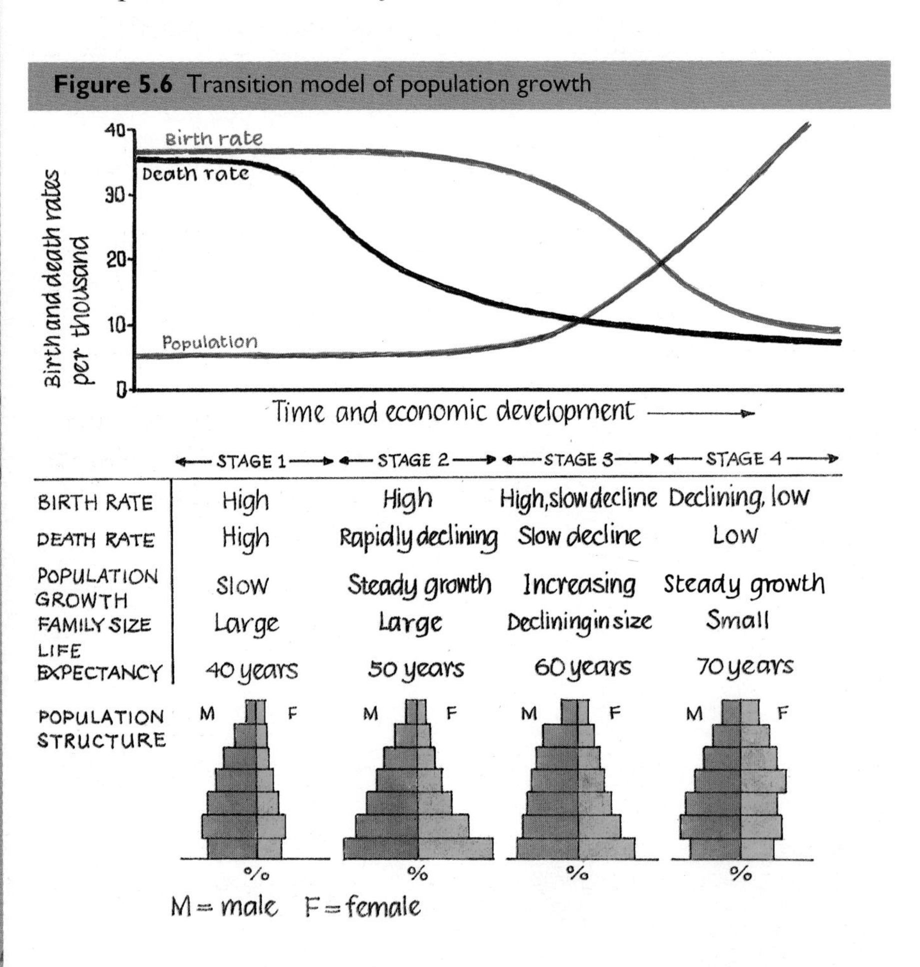

Figure 5.6 Transition model of population growth

Population structures

The changes to a population structure can be shown using age–sex diagrams. These give the number of each sex for each age group. With a rapid population growth there will be a broad base to show the large number of young people (Figure 5.7a). In countries with a low population growth the base is narrow (Figure 5.7b).

People too young or old to work are called dependants. These people depend on those working to support them. By dividing the number of dependent by those working, a **dependency ratio** can be calculated. The **dependency ratio** tends to be higher in MEDCs, i.e. 1 to 3, than in LEDCs, i.e. 1 to 6. In LEDCs most dependants are children but in MEDCs there are more older people who are dependent. Those working, directly support their families and also pay taxes which are used to provide the health and social services for the population.

Figure 5.7 Age-sex pyramids in a) India and b) the UK

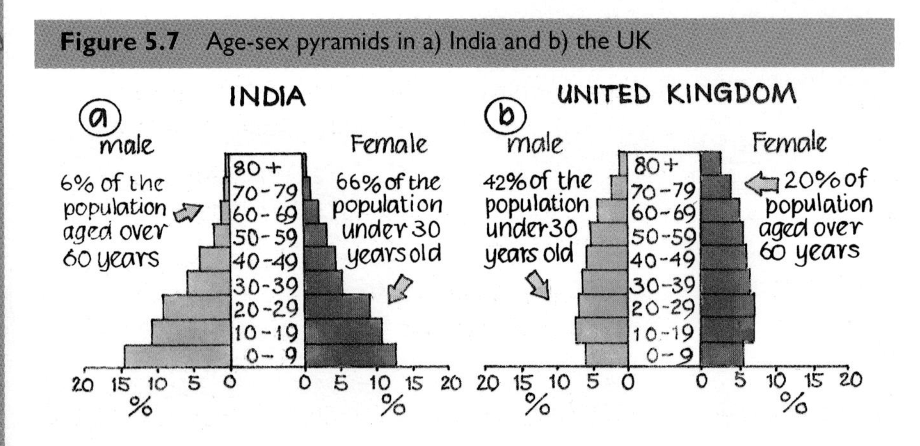

Figure 5.8 Population data

Country	Population in millions	GNP/Person US$	Birth rate (per 1000)	Death rate (per 1000)	Annual population growth (%)	Infant mortality (per 1000)	Number per doctor	Life expectancy (yrs)	Population doubling time (yrs)
Brazil	159.5	3640	20	9	1.1	53	1000	60	40
France	58.8	24 990	13	9	0.4	6	333	79	182
Germany	82.2	27 500	9	11	−0.2	6	370	76	–
India	980	340	25	9	1.8	69	2439	60	32
Japan	125.9	39 640	10	8	0.2	4	600	80	250
Mali	11	250	51	19	3.2	101	20 000	48	21
Mexico	97.4	3320	26	5	2.1	24	621	74	32
Nigeria	88.5	260	43	12	3.1	70	5882	53	23
UK	58.3	18 700	13	11	0.2	6	300	76	250
USA	263.6	26 980	15	9	0.4	7	420	76	125

Comparing populations

Differences between LEDCs and MEDCs population are due to social and economic factors. Figure 5.8 suggests thet LEDCs have a lower life expectancy, a higher infant mortality and more people per doctors than MEDCs. The low incomes mean LEDCs are unable to spend large amounts of money on healthcare provision.

The average family size is smaller in MEDCs. The population growth rate is very low compared to LEDCs. A high birth rate and falling death rate mean that LEDC populations are doubling within 20–30 years. LEDC children spend less time in education and begin working much younger than in an MEDC.

Word box

birth rate The number of babies born per thousand of the population each year

death rate The number of people who die per thousand of the population each year

standard of living How well the population's basic needs, food, water, health care and shelter are met

dependency ratio The ratio of people working to support dependants such as children or old people to the rest of the population

QUESTION BOX

1 When would a population increase?

2 Draw a copy of the Transition Model. Label stages 1, 2, 3 and 4.

3 State how the birth rate and death rate change at each of the four stages.

A Explain how changes to the birth rate and death rate affects the population structure.

B Using Figure 5.8, suggest a Transition Model stage for each country. Explain how you decided the stage for each country.

C How are the population indicators for an MEDC different from those of an LEDC?

Extension task: see worksheet 5A/B

Why do people move?

People move for a variety of reasons. Many move to improve their standard of living or in search of work. Others are forced to move from their homes for different reasons. They may leave because they fear persecution on account of their race, belief, nationality or social group. People who leave their home country to go and live permanently in another country are known as **emigrants** (see pages 58–59).

As people become more aware of other places and as transport has improved, many countries have seen large movements in population. The movement of people from rural areas to urban areas has occurred in many countries. People who want to escape rural poverty are attracted to urban areas. Young people are more able to move from rural areas than the older generations who have settled in an area.

Figure 5.10 Migration in Brazil a) Farmed scrubland, Bahia b) Ouro Preto c) São Paulo

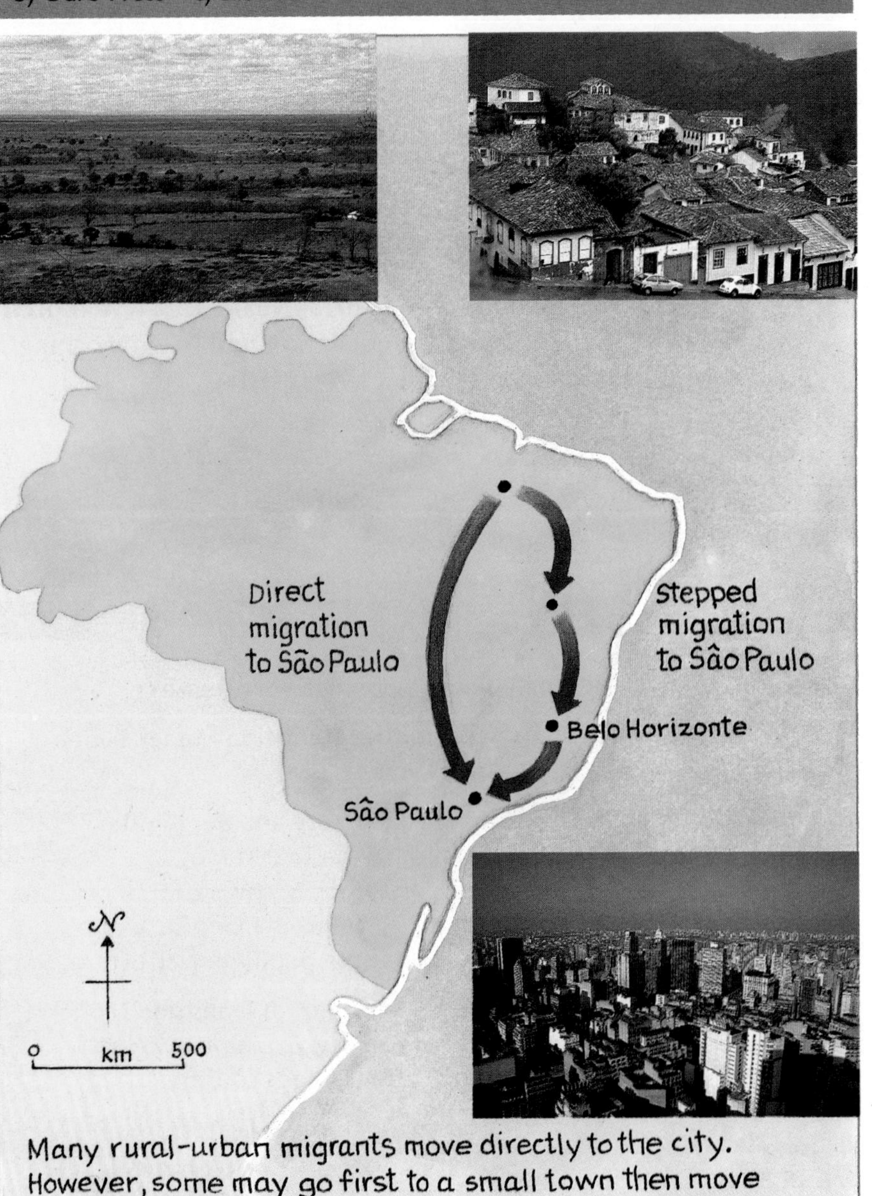

Population movement in Brazil

In Brazil the population has grown rapidly (Figure 5.14). The greatest population growth has been along the Atlantic coast. Few people live in the interior of Brazil because of travel difficulties and its isolation. Farming is the only way to earn a living. There are large deposits of raw materials but the lack of roads and railways means that workers cannot travel to the area and the materials cannot be transported away: so much of the interior is **underdeveloped** (see pages 96–97).

Word box

emigrant A person who moves to live in another country

underdeveloped Describes a poor region with few industries

favela An unplanned area of poor housing built from available materials

Figure 5.11 Cattle ranch at Bahia, Brazil

Figure 5.13 Population density in Brazil

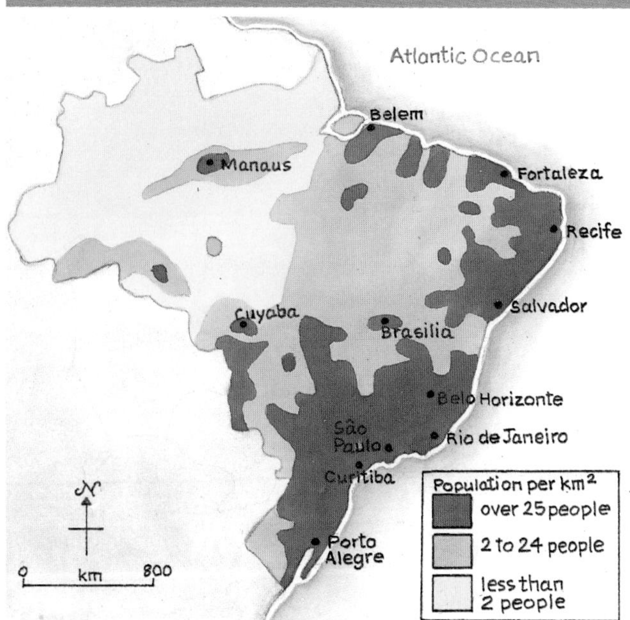

Many areas in the interior of Brazil are very poor. There are no paid factory jobs. Farming is a hard and often uncertain way to earn a living. The cities seem to offer the prospect of work and a better standard of living, but those who migrate to urban areas find they swap one type of poverty for another. The cities are crowded. Cities like Sao Paulo do not have enough affordable housing for newcomers to rent or buy; instead large areas of shanty towns or **favelas** have been built. São Paulo has large areas of shanty towns or **favelas** (Figure 5.12). These are built from whatever materials are available, lack basic services are built near industry, and often on hillsides. In addition the favelas are often illegal. The inhabitants also face many other problems, for example a lack of schools, rubbish collection and medical services. Life in the favelas is difficult and unhealthy.

Figure 5.14 Population growth in Brazil

	1860	1880	1900	1920	1940	1960	1980	2000	2025
Population in millions	9	11	18	28	39	72	121	179	199 (projected)
Percentage of population living in urban areas	–	–	–	12	31	45	78	83	85

QUESTION BOX

1 Use Figure 5.14:

a) Plot as proportional lines the rural and urban populations for 1940, 1980 and 2025.

b) Describe how the numbers living in urban areas have changed.

2 Draw a sketch of the houses in Figure 5.12. Describe what you think living here might be like.

A Explain why there has been a movement of people from rural to urban areas.

B Explain why an increase in Brazil's urban population is likely to cause problems in the future.

Figure 5.12 Favelas, Rio de Janeiro, Brazil

International movement

West Indians come to Britain

Since 1945, there has been an increase in international **migration**. There has been a movement of people from less developed countries to the richer, industrialised nations. These migrants moved to find better paid work.

One group who were encouraged to come to Britain were West Indians. Britain and other European countries needed workers to help with the post-war economic recovery.

Figure 5.15 Montego Bay, Jamaica

At the time the West Indies had an unemployment problem. Many West Indians had looked for work in the USA, but the USA had passed the McCarren-Walter Act which limited West Indian immigration. The British Government claimed that their Welfare State offered full employment to all and so thousands of unemployed West Indians made the long trek to Britain to find work, even if this meant accepting low wages. Companies like London Transport and the National Health Service advertised to recruit people from the West Indies.

Figure 5.16 Immigrants from Jamaica arrive at Tilbury Docks on SS Windrush in 1948

Figure 5.17 Number of West Indian immigrants entering Britain

Year	Number
1958	15 000
1959	16 400
1960	49 600
1961	66 300
1962	34 000
1963	7900
1964	14 800
1965	14 800
1966	10 900
1967	12 400
1968	7000
1969	4500
1970	3900
1971	2700
1972	2400
1973	2600
1974	3100
1975	3600
1976	2600
1977	2200
1978	1700
1979	1200
1980	1000
1981	900
1982	700
1983	700
1984	600

Restrictions on immigration

The large number of immigrants from the West Indies, coming to Britain in the 1960s, were joined by other people from the **New Commonwealth** and those escaping from civil wars and persecution. The British Government came under pressure to restrict the numbers entering the country. Several laws were passed to reduce the numbers. These included the following.

1948 The British Nationality Act gave all citizens of the dependencies and Commonwealth right of entry to Britain.
1962 The Commonwealth Immigrants Act restricted immigration to those who had UK passports, a Ministry of Labour work voucher or a skill which Britain needed.
1968 The Commonwealth Immigration Act distinguished between citizens who had a grandparent or parent born in the UK and those who had not, only the former being allowed entry.
1971 The Immigration Act restricted immigration for settlement to wives or husbands and the dependants of those who had settled in the UK for five years.
1982 The British Nationality Act gave people with British Citizenship only the automatic right to settle in the UK.

The laws changed the basis for entry into Britain away from economic and towards family criteria. Black immigrants faced varying amounts of racial discrimination.

Many immigrants chose to live in the inner suburbs of towns and cities. Here there was cheap housing near factories and many took low paid manual work. By living together immigrants were able to keep a **cultural identity**, a sense of community and a feeling of security.

Many of the immigrants who came looking for work in Britain stayed and had families. This has resulted in a second generation of immigrants who have become part of British society with many becoming successful business people, making an economic contribution to the country.

Figure 5.18 Integration in the playground

Word box

migration The movement of people from one region or country to another
New Commonwealth This includes the countries of India, the West Indies and East Africa
cultural identity A shared way of life, set of beliefs and language

QUESTION BOX

1 Why did the British Government encourage West Indians to come to Britain?
2 Suggest why Britain was attractive to West Indian immigrants?
3 What problems do you think immigrants might face when moving to another country?

A Using Figure 5.17, draw a line graph to describe how changes in the legislation affected the numbers migrating into Britain.
B If you had to emigrate where would you go, why, and what problems do you think you would have?

Extension task: see worksheet 5C/D

Future problems

The world's population of 5.7 billion could reach over 10 billion by 2025. The potential problems caused by this increase were discussed at the Cairo Conference on Population in August 1994. Experts fear that many more people will be living in poverty, that there will be further damage to the environment with conflict over scarce resources such as water and farmland. The greatest growth of population is likely to be in the poorest countries, although better health care has improved life expectancy and survival world-wide.

Developed countries

By 2040, one in every four people in Europe will be over 65 years old (Figure 5.19). In richer countries there has been a decline in the number of babies being born. In Britain on average, each woman needs to have 2.1 children to maintain the population. At present the actual figure is 1.9 children and this trend of having less children is common in richer countries. The reason for declining populations is not **contraception** but improved living standards. Smaller families means that there is more money to be spent on consumer and luxury goods.

Countries with falling populations, an ageing workforce and high living standards encourage immigrants to come and work, usually for low wages. However, the presence of large numbers of immigrant workers can cause social problems and they may feel resentment at being viewed as second class citizens (Figure 5.20).

Developing countries

To feed the expanding population in developing countries an increase of 56 per cent in arable land is needed by 2050. This will destroy many irreplaceable natural wild areas like forests and wetlands. Poor farming methods and **desertification** are destroying marginal farmland in areas like the Sahel in Africa (page 42). The world's renewable resources like firewood are used faster than they are replaced. The benefits of **industrialisation** in the developing world are

Figure 5.19 The greying population of the UK

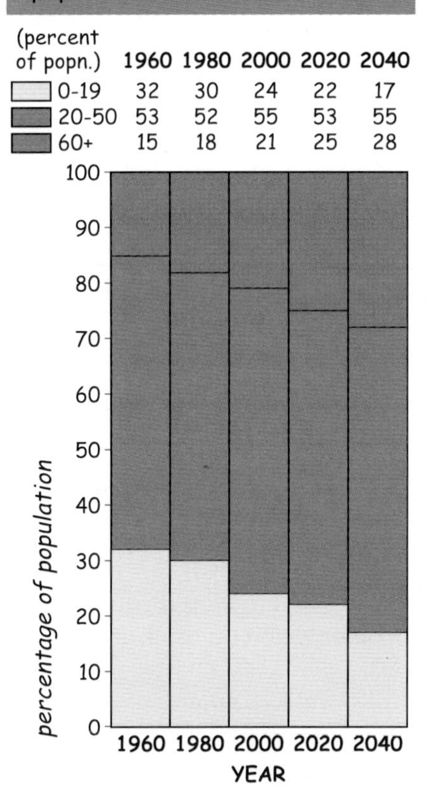

(percent of popn.)	1960	1980	2000	2020	2040
0-19	32	30	24	22	17
20-50	53	52	55	53	55
60+	15	18	21	25	28

Figure 5.20 Racial tensions in Germany

Figure 5.21 Family planning poster, Szechwan, China

Figure 5.22 Industrial pollution in Chelyabinsk, CIS

reduced by the damage from environmental pollution. In the LDCs, as a result, there is likely to be a decline in living standards with 1156 million people in poverty.

Solutions

Many population planning policies have emphasised the need for contraception. This approach has been opposed by various social and religious groups who are against **family planning**. In China, the policy of one child per family is related to family allowances and privileges and there is a great pressure to conform (Figure 5.20). In many Catholic South American countries the Church actively campaigns against family planning.

A United Nations Report suggests that the best way to control the population explosion in the developing world is through education. Literacy levels are high in MDCs and lower in LDCs, especially for women. In many societies women have a low status and their contribution to the family is undervalued. If women were better educated and listened to, they could make an informed choice about whether to have smaller families. For instance, in Zimbabwe, women with no formal education have average family sizes of seven children, but educated women have only four children.

The hope that further industrialisation in developing countries will improve their population's standard of living is unlikely to be realised. Trade between developed and developing countries still favours the richer, industrialised countries. The use of self-help and small-scale projects such as textile co-operatives in India could be a valuable strategy. The Rio Earth Summit in 1992, suggested that improvements should be based on **sustainable development**. This may not bring about the standard of living enjoyed in the developed nations but it could provide the basis for stabilising the population.

Word box

contraception Any form of birth control

desertification The spread of deserts caused by poor farming methods

industrialisation The development of manufacturing industry

family planning Help and advice about contraception

sustainable development Where resources are replaced after being used

QUESTION BOX

1 Write a list of possible problems caused by the continued increase in world population. Compare your ideas with those of others. Which do you think are the most serious problems and why?

2 Suggest how population growth could be reduced.

A Explain why improving women's rights and education could reduce the number of children they have?

B Using Figure 5.19 explain what is meant by a 'Greying Europe'. Why is this population trend a problem?

SUMMARY QUESTIONS

1 Using named examples, describe why some areas have a sparse population and others a dense population.

2 Describe how birth rate and death rate can change over time.

3 Suggest why people choose to emigrate from LEDCs to MEDCs.

A Explain how a natural region can affect the population density?

B Explain why people decide to move to another region or country.

C Explain why improving the standard of living can help reduce population growth.

unit 6

A Place To Live

Key questions

- What factors influence the location of a settlement?
- Which regions are the most urbanised?
- Is there a link between settlement size and range of services?
- How different are parts of an urban area?
- What are the similarities and differences between settlements in MEDCs and LEDCs?
- What causes counter–urbanisation in MEDC settlements?

Settlement sites

The earliest recorded settlements are found in the Indus Valley and the Middle East. These are over 4000 years old. The **site** chosen for a settlement depends on the quality of the land and its position in the area relative to other features and amenities (Figure 6.1). A site may take advantage of the natural features such as water supply or high land safe from attack. Many early settlements were sited on natural routeways (Figure 6.2).

As trade increased, small settlements grew into market towns. Settlements which were located near raw materials became industrial centres. These industrial towns grew quickly as they attracted better transport links and other industries and workers to the area.

Figure 6.1 Settlement site factors

Near other settlements?

Raw materials for buildings or industry?

Flat land for building?

Woodland nearby?

Can the site be defended?

Fertile farmland?

Land for grazing animals?

Reliable water supply such as a well or river

Natural routeways – a river?

Can the river be bridged or forded?

Can boats sail up from the sea?

Is there a road?

Urbanisation

Urbanisation in the UK began after the start of the Industrial Revolution in the eighteenth century. The early factories attracted workers from rural areas and the urban population in the industrial towns grew steadily. During the past 100 years, many small settlements have grown and merged together forming a continuous urban area. These vast urban areas are called **conurbations**, e.g. Manchester and London (Figure 6.3).

Urban growth is a world-wide feature (Figure 6.4). In 1920 only six per cent of people in the developing world lived in towns of over 20 000 people; today it is over 35 per cent. This rapid urban growth in the developing countries has caused a number of problems, e.g. overcrowding, traffic congestion and insufficient services like water supply and sewerage. Many developing countries are experiencing rapid urban growth as people leave the countryside to find work in the cities.

In developed countries people are now moving out of the inner city to live on the urban fringe or in the countryside. Here they hope to find a better residential environment. People are able to commute to work using public and private transport.

Figure 6.2 Examples of settlement sites

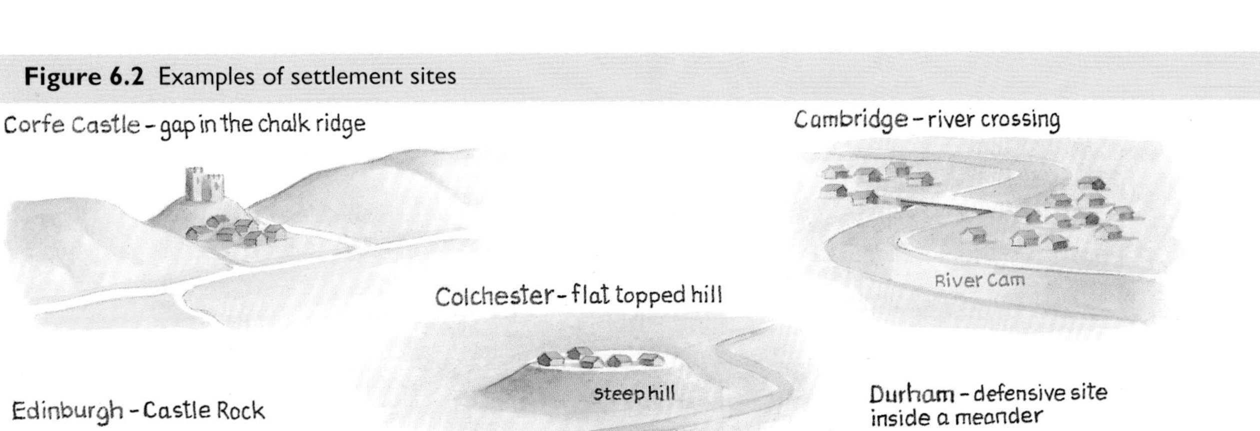

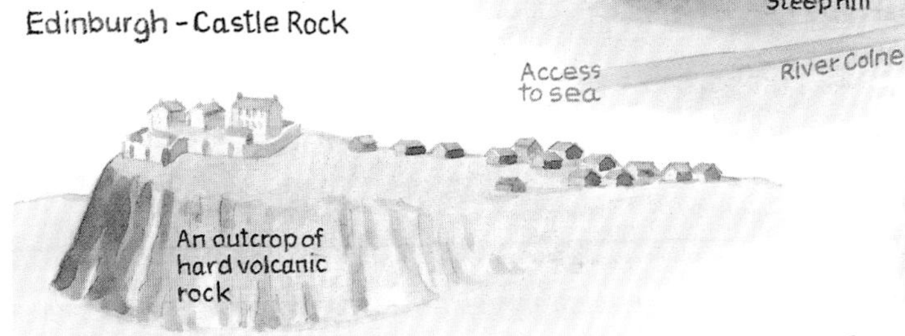

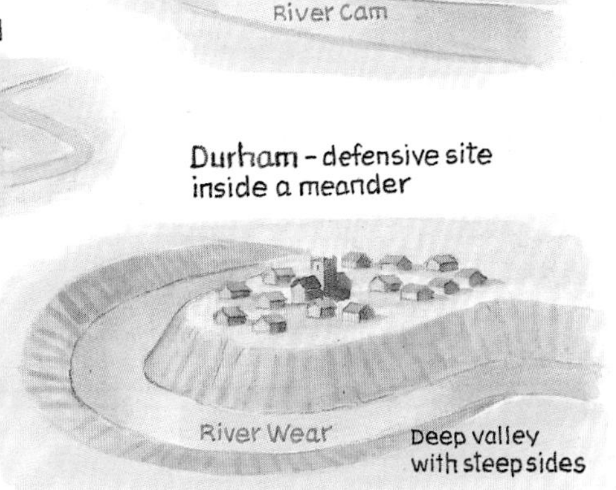

Figure 6.3 London's growth

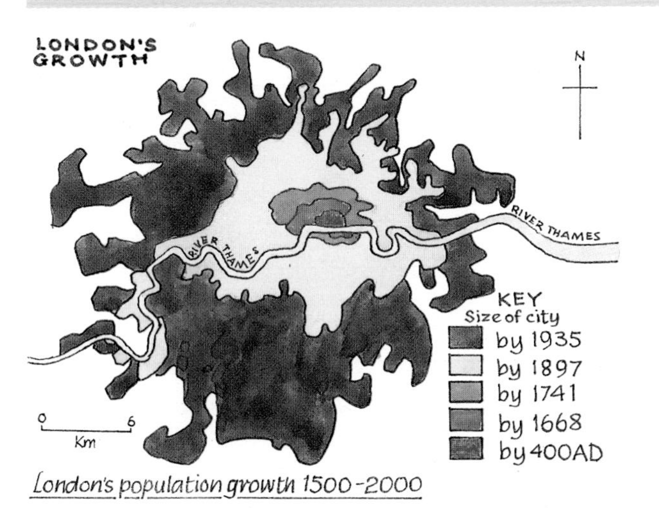

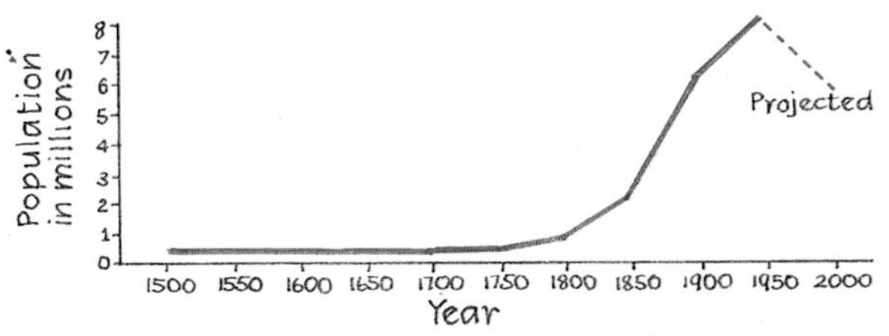

Figure 6.4 World urban population

The percentage of the population living in urban areas, 1950–2000

Region	1950	1970	1990	2010 (projected)
Africa	15	23	31	44
Latin America	41	57	71	78
North America	64	70	75	82
Asia	16	25	32	48
Europe	54	64	73	79
Russia (CIS)	39	56	74	84
Oceania	61	70	70	84

QUESTION BOX

1. Use Figure 6.2 to identify a reason for the siting of each settlement.
2. Using Figure 6.4, plot Africa and North America as a line graph. Describe the difference between these two regions.

A Suggest reasons for the location and site of local settlements where you live.

B Use Figure 6.4 and a blank world outline:

a) plot using graded shading the urban population for 1950, and 2010.

b) calculate the change in the urban population, 1950–2010, for each region.

c) describe the trends in world urbanisation.

Extension task: see worksheet 6A/B

Word box

site The land on which a settlement is built

urbanisation An increase in the percentage of people who live in towns and cities

conurbation A large urban settlement which is the result of towns and cities spreading out and merging together

Settlement functions

Settlements have a variety of purposes or **functions** (Figure 6.5). The function may be related to the main type of industry found in the settlement or related to why a particular settlement is important. Some functions are common to most settlements, for example housing or every day shops and services. However, the larger a settlement, the more functions and greater range of services it will have. This link between the size of the settlement and the number of functions and services produces a **settlement hierarchy** (Figure 6.6).

Shops and services can be described as **low** or **high order goods** and services (Figure 6.7). Low order services include every day shops such as greengrocers or newsagents. These shops need a small number of regular, local customers in order to trade. This type of shop is found in smaller settlements and the suburbs of larger towns and cities. High order goods are more specialised and are bought less often. These are described as consumer goods and include expensive items like jewellery, televisions and furniture. Specialised services like solicitors, entertainment and health care are also high order services. These services need a large catchment area of people to support them.

Figure 6.5 Functions of a settlement, Chesterfield in Derbyshire

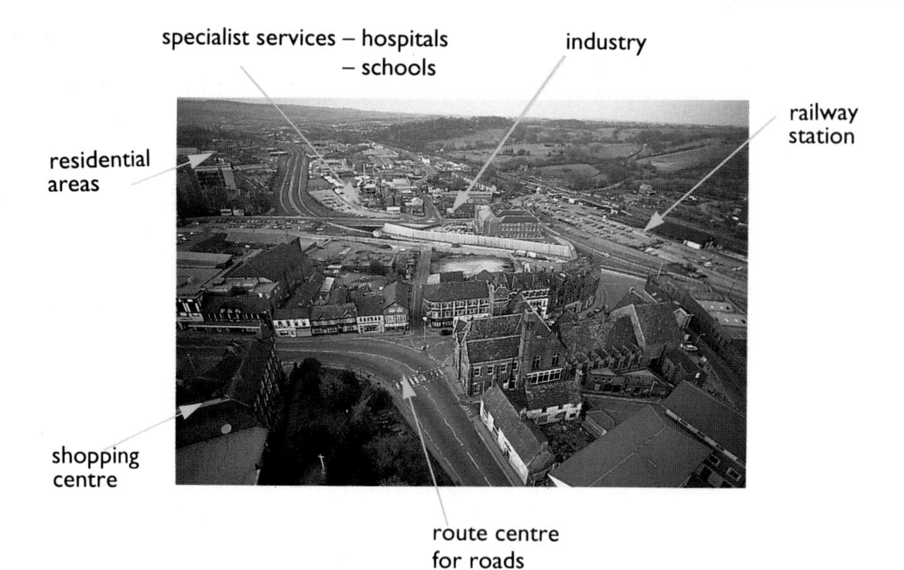

Figure 6.6 Settlement services linked to population

site The land on which a settlement is built
urbanisation A n increase in the percentage of people who live in towns and cities
conurbation A large urban settlement

which is the result of towns and cities spreading out and merging together**Large population**

2 million people
500 000 people
100 000 people
10 000 people
2000 people
a family

Small population

Figure 6.7 Examples of shops and services

Examples of high order shops and services – usually found in the town centre

- Specialist shops, e.g. electrical items, jewellery, records, clothes, shoes
- Large department stores
- Library
- Town Hall; local government offices
- Financial services, e.g. banks, building societies, insurance
- Solicitors and legal services
- Entertainment

Examples of low order shops and services – usually found in the suburbs

- Every day food shops, e.g. greengrocers, bakers, butchers
- Small supermarket
- Take away food
- Post Office
- Hairdressers

Figure 6.8 High and low order catchment areas

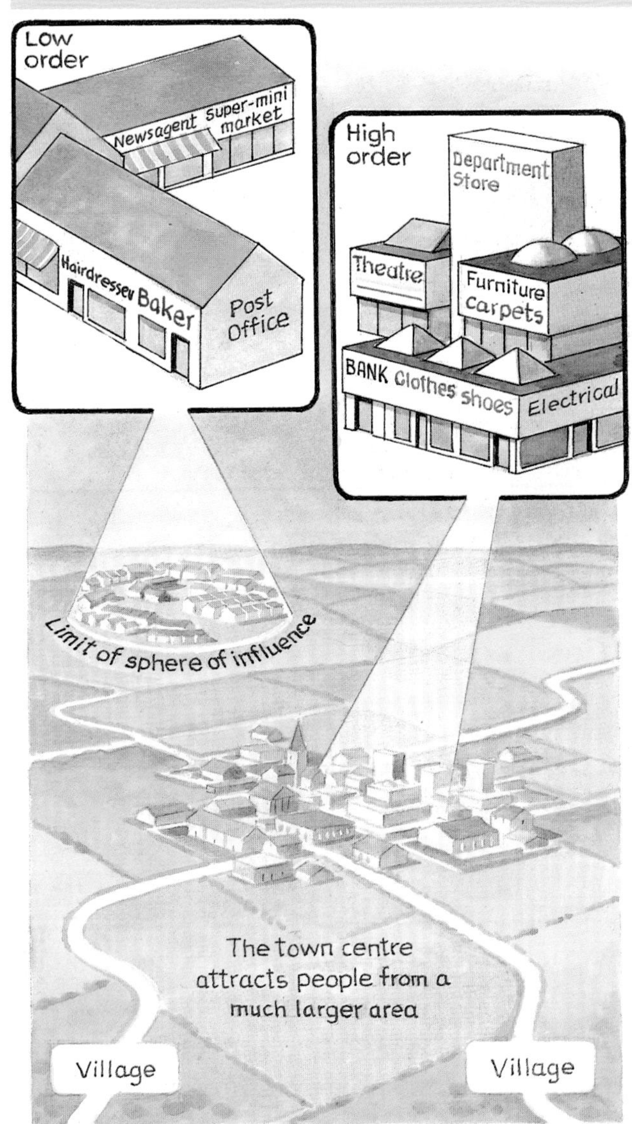

The distance people are prepared to travel to buy a good or use a service in a catchment area is called the **sphere of influence** (Figure 6.8).

The pattern of shops and services is changing with the development of out of town centres, e.g. Meadowhall in Sheffield. Most towns now have large supermarkets or shopping complexes built on spacious sites at the edge of the town. Land is cheaper here and there is easy access for customers who travel by car. These centres are modern and contain a wide range of under cover shops and services with large free car parks.

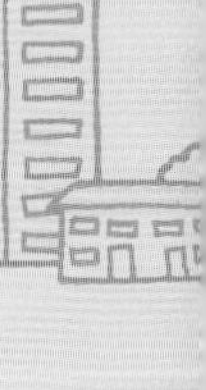

Despite the attractions of out of town centres, they have changed the way people shop and have taken people and trade away from the town centre.

QUESTION BOX

1 List the name and type of shop or service found locally. Describe the range of goods and services available.

2 Plan a Shopping Habits Survey to test this idea, 'People frequently use local shops to buy everyday items.' These questions will help:

- which local shops do you use, why, and how often?
- where do you go for your main food shop, why, and how often?
- where do you go to buy consumer goods, why, and how often?

Can you think of any other questions? Ask at least ten people these questions. Present your results as a summary table, then graph and describe your results.

A Explain why the range of services available depends on the size of the settlement.

B Newspaper adverts and reports provide a useful way to find out about a settlement's Sphere of Influence. Use a local map and plot on the place names mentioned in a newspaper article. Draw a line to connect the outer points. This represents a possible Sphere of Influence. Describe your completed map and suggest how its accuracy could be improved.

Word box

functions What makes a settlement important
settlement hierarchy An order of settlements according to population size and the range of goods and services available
low order good A low value, every day commodity which can be bought locally, e.g. newspapers
high order good A higher value consumer item or service found in a specialist shop or office in a town centre
sphere of influence An area served by a shop or service

Developed and developing cities

Cities in developing countries: case study São Paulo, Brazil

Many of the migrants who make up São Paulo's population arrived in the city hoping for a job and a better standard of living. These incentives are known as **pull factors**. **Push factors** are those which make people want to leave an area or country, e.g. poverty, unemployment (Figure 6.9).

However, the urban areas in São Paulo are overcrowded and there has been a lack of plans to cope with the increase in people. Sprawling areas of squatter housing or **favelas**, have been built on the edge of São Paulo, standing in sharp contrast to the modern city skyline.

Figure 6.9 Push-pull factors

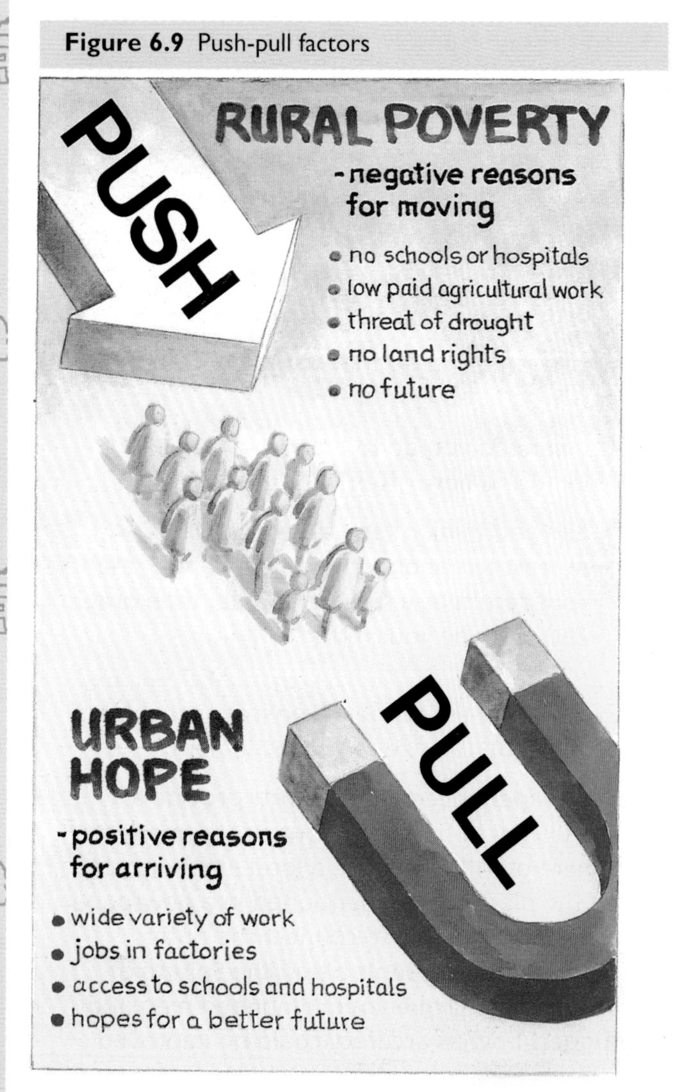

More than six million people live in the favelas (Figure 6.10). São Paulo has many problems which affect the quality of life of its inhabitants, e.g. pollution, traffic congestion and poor housing. There are plans to improve the favelas by providing water, sewers and electricity.

São Paulo is an important industrial centre and several foreign companies have bases here, e.g. Fiat, Volkswagen and Ford. The migrants provide a cheap and plentiful workforce. Although the jobs available are low paid, the migrants can still earn more than in the countryside and they have access to a greater range of services.

Cities in developed countries: case study Rome, Italy

Rome is the capital and administrative centre of Italy. It did not develop as an industrial area. It is a major tourist attraction. The Roman Catholic Church helped to rebuild Rome from 1500–1700 and the Vatican City is the headquarters for the Catholic Church and the residence of the Pope.

As a city grows it develops certain patterns of land use with particular characteristics. These can be mapped to show a number of urban areas (Figure 6.11). Rome has both a **central business district** and an historic centre,

Figure 6.10 São Paulo's location

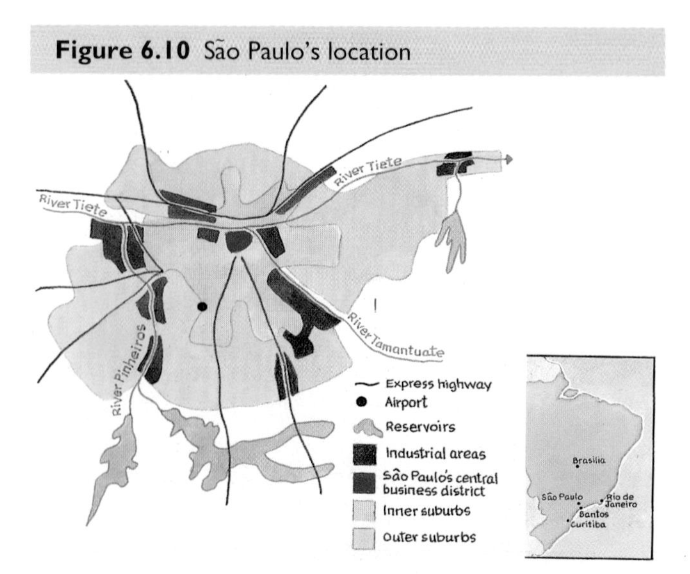

very close together. Early growth was close to this central area and included a mixture of industry, housing and open space. These formed the **inner suburbs**. As public and private transport became widespread, the city grew outwards. people were able to live further away from the city centre and **commute** to work. The **outer suburbs** have large residential areas separated from industry by open spaces (Figure 6.12). Some areas of the inner suburbs became run down and changed into unattractive **slum areas**, housing migrants. Rome's population doubled from 1950–80 as migrants from the south looking for work arrived. There were not enough houses and 300 000 people still live in illegally built homes. Plans to solve traffic congestion and reduce air pollution which threaten ancient monuments have failed. There are also social problems, e.g. unemployment, crime and racism.

QUESTION BOX

1 Describe the urban problems found in São Paulo and Rome. Explain why some problems are common to both cities.

2 Describe, using diagrams, how Rome has grown.

Figure 6.11 Concentric model of Rome

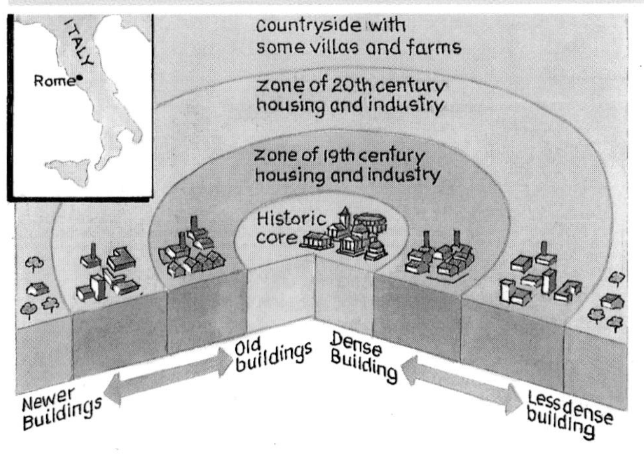

QUESTION BOX

1 What is meant by 'push–pull' factors. Give some examples of each.

2 Use Figure 6.11 to describe how the urban area would change from the centre of Rome to its rural fringe.

A Has Rome's urban area grown in the same way as Sao Paulo's?

B Using Figure 6.12, describe the street layout, buildings, land use and population density of the two areas shown.

Figure 6.12 Street patterns in Rome, photos a) Central Rome, River Tiber b) St Peter's Square and view of Rome

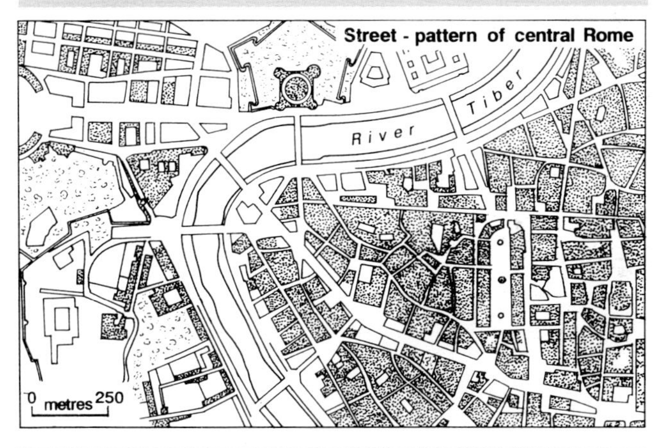

Urban problems in the UK

Cities in developed countries experience a number of different problems to those in LDCs. These include traffic congestion, slum housing, **derelict land** and social problems. There has been a steady decline in the population of the inner suburbs (Figure 6.13).

Figure 6.13 Population change in Derby, 1971–2001

Percentage of the population

Derby	1971	1976	1981	1986	1991	1996	2001
Inner suburbs	41	38	34	31	30	28	27
Outer suburbs	59	62	66	69	70	72	73

People have left the older and run down areas to live in the outer suburbs. Many people have moved to find a better residential environment. The old terraced housing and factories found in the inner suburbs are being replaced by large office buildings, high rise housing, car parks and ring roads. However, not all the residents are able to leave, the old and those who cannot afford to move remain. There are also large numbers of bed-sit flats where residents stay for a short time and then move on (Figure 6.14 and 6.15).

There is concern about the growth of the outer suburbs. Many people feel that the land in the inner city should be put to better use. The new residential estates take valuable agricultural land and spoil the landscape. Local villagers are unable to compete with commuters with higher incomes when buying or renting property.

Efforts have been made to improve the inner suburbs. Local Authority grants have helped to improve housing. Slums have been replaced by high rise housing which are, however, often worse. **New Towns**, e.g. Telford and Harlow, were planned to provide a good residential environment with work and services close by. Planners have created **green belts** around cities to control urban growth and stop the merging of towns and cities.

Figure 6.14 Derelict buildings, London

Figure 6.15 Modern housing estate. Reading

Figure 6.16 Form for environmental study

At each point, look around and score the area out of ten for each feature

Category	10	0	1	2	3	4	5	6	7	8	9	10
Litter	Clean streets, no litter	Litter scattered everywhere										
Pavement	New, level, smooth	Cracked paving, ruts and holes										
Advertisement	None	Several posters pasted to walls										
Car parking	No cars parked on street	Several cars parked on streets										
Sky	No overhead wires	Sky criss-crossed with wires										
Appearance	Bright, lots of vegetation	Dark, little vegetation										
Condition of buildings	Well-maintained	Poorly maintained, buildings derelict										
Road safety	Wide pavements, no obstructions	No pavements, several obstructions										
Air freshness	Fresh	Smell of pollution										
Noise	Silence	Constant road or industrial noise										

0 means that it is very poor.
A score between 1 and 4 means that the area is poor.
5 means that the area is satisfactory.
A score between 6 and 9 means the area is good.
A score of 10 means the area is excellent.

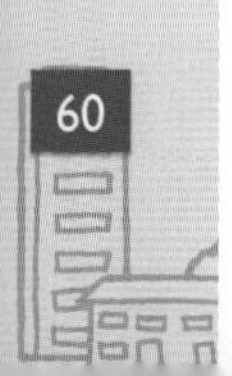

Figure 6.17 Field sketch

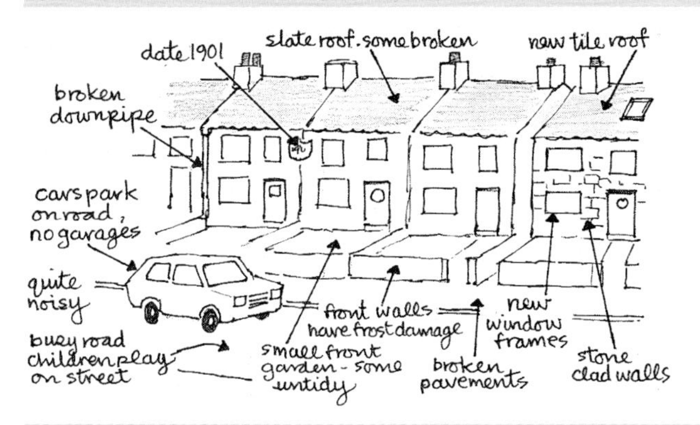

Figure 6.18 Example of multi-line graph for a survey point

Category	0 1 2 3 4 5 6 7 8 9 10
Litter	4
Floorscape	5
Advertisements	4
Car parking	3
Wirescape	8
Landscape and vegetation	7
Conditions of buildings	8
Road safety	4
Air freshness	9
Noise level	5

QUESTION BOX

1. List five problems found in the inner city.
2. Compare Figures 6.14 and 6.15:
   a) describe the appearance of each area.
   b) which do you think is the best area.
   c) give a reason for your answer.
3. Explain one way an urban area could be improved.

A Use Figure 6.13:

a) plot, as a line graph, changes in the inner and outer city populations.

b) describe the trend shown.

c) suggest some reasons for this trend.

B Explain why counter–urbanisation is a problem for cities like Derby.

Word box

derelict land Industrial waste land

New Towns A planned new, self-contained, urban area built away from existing cities

green belt An area of land around a city or town on which there is a control on urban development

SUMMARY QUESTIONS

1 Explain what is meant by the term urbanisation.

2 Suggest some factors which influence the siting of a settlement.

3 What is meant by a settlement hierarchy?

4 What are the advantages of shopping in a large out of town centre?

5 Describe some of the urban problems found in cities like Sao Paulo.

6 Draw a labelled sketch of the street where you live.

7 Write a list of all the good points and a list of the problems of living where you do.

8 Suggest some ways to improve your street.

A Describe the pattern of global urbanisation.

B Explain why you would find more newsagents than electrical stores in a city.

C What urban features would you expect to find in Rome and Sao Paulo? Which urban features would you only expect to find in Sao Paulo?

D Design an illustrated Visitors Guide for a walk around your local area. Include a street map showing the route, fieldsketches and a description of at least five stops.

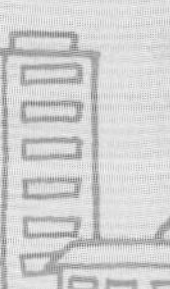

unit 7

JAPAN

Key questions

- Where is Japan and what is it like?
- What are the different regions of Japan like?
- What are Japan's economic links with the rest of the world?
- How and why is Japan changing?

Introduction

Figure 7.1 Japan: location map

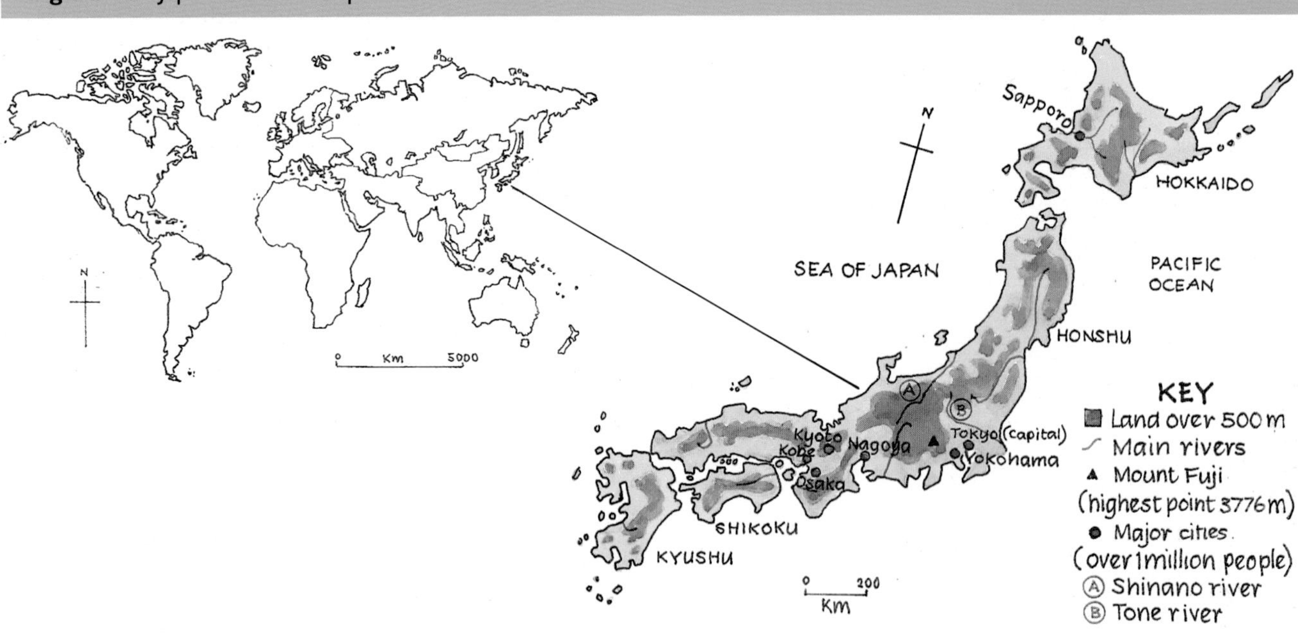

Japan is made up of four main islands and nearly 4000 smaller ones (Figure 7.1). Much of the country is mountainous with most of the flat land near the coast. The climate varies from warm in the south to cold in the north. It has more than its fair share of **natural hazards**: there are over 60 active volcanoes; it has, on average, 7000 earthquakes a year; **tsunamis** affect its coastline; and between June and October it is in the path of **typhoons**.

Japan has a population of 125.9 million (1997) and is the seventh most crowded country in the world. Seventy-six per cent of its population live in towns and cities, the largest of which is the capital, Tokyo, with eight million inhabitants. Most Japanese enjoy a high standard of living; their country is very rich (**GNP per person** is one of the highest in the world) and it has the highest **life expectancy** of any in the world (Figure 7.5). However, Japan also has problems such as traffic congestion and industrial pollution.

There are big differences between the old and the new in Japan. Traditional costumes, the tea ceremony (Figure 7.2) and martial arts like karate, are part of the old ways. Western-style clothes, electronic gadgets and McDonalds are part of the new. Things have changed so much in the last 50 years that many older people in Japan find it difficult to get along with the younger generation.

Figure 7.2 English pupils are taught about the tea ceremony as part of a 'Japan Day' held at their school

Figure 7.3 Daisetsuzan Mountains in Hokkaido, Japan

Word box

natural hazard A violent natural event which threatens people
tsunamis Sea waves caused by earthquakes
typhoons Violent tropical storms (also known as hurricanes)
GNP per person The amount a country earns divided by its population
life expectancy How long, on average, people are expected to live
relief The height and shape of the land

Figure 7.4 Industry on Tokyo Bay, Japan

Figure 7.5 Fact file: Japan, the UK and India

	Japan	UK	India
Population (millions) 1997	125.9	58.6	980
Area (km^2)	327 800	244 880	3 287 000
GNP (US $ per person) 1995	39 640	18 700	340
Employment structure	primary 7% secondary 34% tertiary 59%	primary 2% secondary 28% tertiary 70%	primary 62% secondary 11% tertiary 27%
Population growth rate (per year)	0.3%	0.3%	2.5%
Life expectancy	80	77	61
Foreign aid (US $ per person	106 given	53 given	2 received
Food supply (calories per person per day) 1992	2903	3317	2395

QUESTION BOX

1 Use an atlas and Figure 7.1.
  a) Name the four main islands of Japan.
  b) Copy and complete this sentence: 'Japan is located in the ______ Ocean to the east of ______, North and South Korea and Russia. It is ______ of the equator and lies between lines of latitude ____° and ____°.'
  c) What is the name of Japan's longest river?
  d) Describe Japan's relief.

2 What can you tell about Japan from each of the photographs in Figures 7.2, 7.3 and 7.4?

A Select and use appropriate techniques to present the information in Figure 7.5.

B Which of the statistics in Figure 7.5 indicate that Japan is more economically developed than India, and why?

Relief and climate

Earthquakes and volcanoes are common in Japan because it lies on the boundary between the Pacific and Eurasian plates (see Figure 1.3, page 5). In fact, nearly all of Japan's mountains are volcanic in origin. At the foot of the mountains, rivers have deposited **alluvial fans**: many of these have been cut into steps (terraces) by farmers for rice fields. Nearer the sea, where the rivers have less energy, they have deposited **silt** to form alluvial plains: these are very important because they provide just about all of Japan's flat, fertile land.

Japan's climate, which can be divided into four zones (Figure 7.6), is affected by three main factors: **air masses**; ocean currents; and relief. Japan's position is where a number of air masses meet. The Siberian air mass brings cold, wet weather to the west of the country in the winter. The polar air mass brings cold, wet air to the north of the country in the winter and the spring. The tropical air mass brings warm, wet air to the south and east of the country in the summer.

The Kuroshio current helps to bring warm weather to much of Japan in the summer; but it also brings typhoons between June and October. The Oyashio current brings cold weather to the north of the country, especially in the winter.

Figure 7.6 Japan's climatic zones

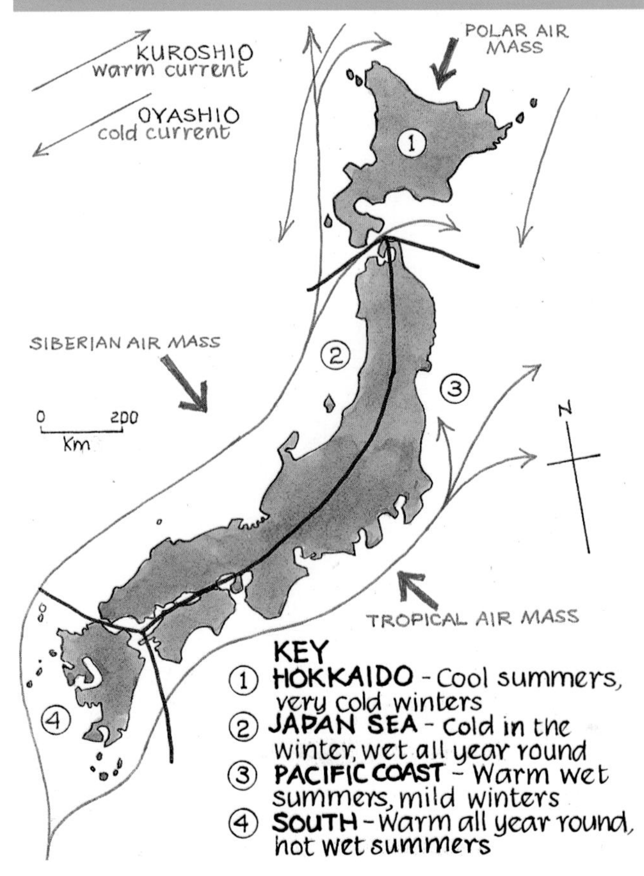

Japan's steep and rugged relief means that there are many **microclimates**; for example, the mountains protect some places from rain but funnel cold winds across others. Also, as you go higher, temperatures fall at a rate of 1°C every 150 m, so it is not unusual to see snow capped peaks even in the warmer south of the country (Figure 7.7).

Figure 7.7 Mount Fuji, Japan

Population and industry

Japan's population has grown quickly this century (Figure 7.8). However, in recent years the rate of increase has fallen and it is expected that by the middle of the next century the number of people living in Japan will start to go down.

Its population is very unevenly distributed (Figure 7.9). The main reason for this is the lack of flat land: 75 per cent of the country has hardly anyone living in it, while average population densities on the coastal plains are 1000 people per km^2.

Figure 7.8 Japan's changing population

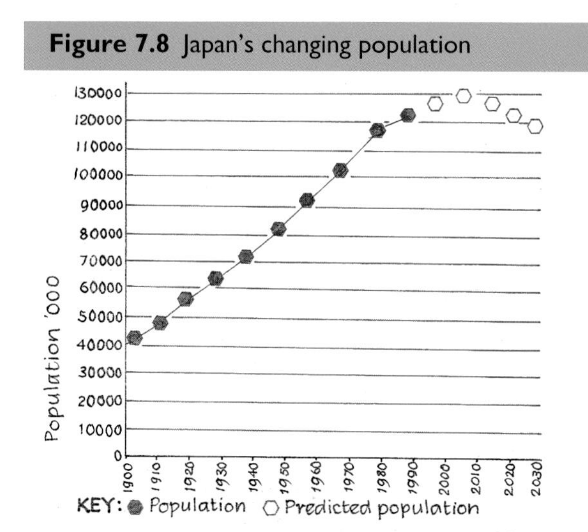

Japan's **occupational structure** has changed as its economy has developed (Figure 7.10). Farming is the most important primary industry, although most farms are small and many of the farmers have part-time jobs in the city. Electronics is Japan's major manufacturing industry, employing 17 per cent of its entire work force; the paper and car industries are also very important. Its tertiary sector has grown because people have more money to spend in shops and on services; banking and international trade have also expanded.

Figure 7.9 Japan's population distribution

Figure 7.10 Occupational structure: 1950 and 1993

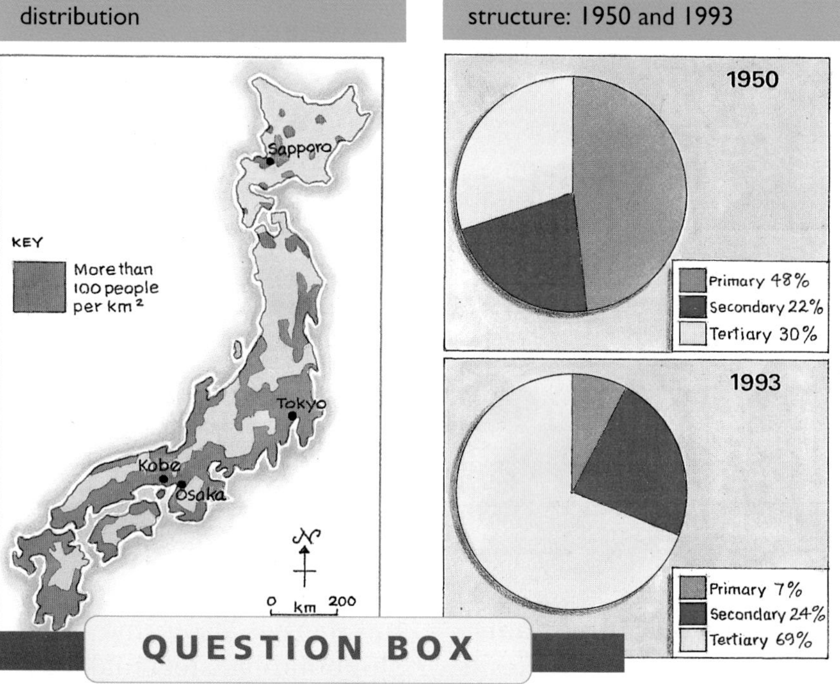

Word box

alluvial fan A steep, fan-shaped deposit of sand and gravel left by a stream at the foot of a mountain range

silt A fine deposit of mud left by a river

air mass A body of air which covers a very large area and which has its own special type of weather

microclimate The climate of a small area caused by local factors, such as relief

occupational structure The balance between primary, secondary and tertiary industry

QUESTION BOX

1 Copy and complete Figure 7.11, 'Japan's climate: a summary'.

2 Why do most people in Japan live near the coast?

3 What has happened to the percentage of people working in primary, secondary and tertiary industry in Japan since 1950?

A Describe and explain a) Japan's main climate zones and b) its population distribution.

B How and why has Japan's occupational structure changed since 1950?

Extension task: see worksheet 7A/B

Figure 7.11 Japan's climate: a summary

	Summer weather	Winter weather	Main air mass	Main current
Hokkaido				
Japan Sea				
Pacific Coast				
South				

Comparing two regions: Hokkaido and Kanto

Japan is divided into eight regions (Figure 7.12), two of which – Hokkaido and Kanto – are compared in this section. In some ways they are similar, for example they are very mountainous and farming is important in both (Figure 7.13). However, in many ways they are different, for example Hokkaido has a much larger area and far fewer people.

Hokkaido

Hokkaido has a harsh physical environment (see Figure 7.3, page 63). It is very mountainous and the winters are very cold (Figure 7.14), and this helps to explain its low population density – it is, in fact, the largest region in Japan but it has the lowest population.

Figure 7.12 Japan's regions

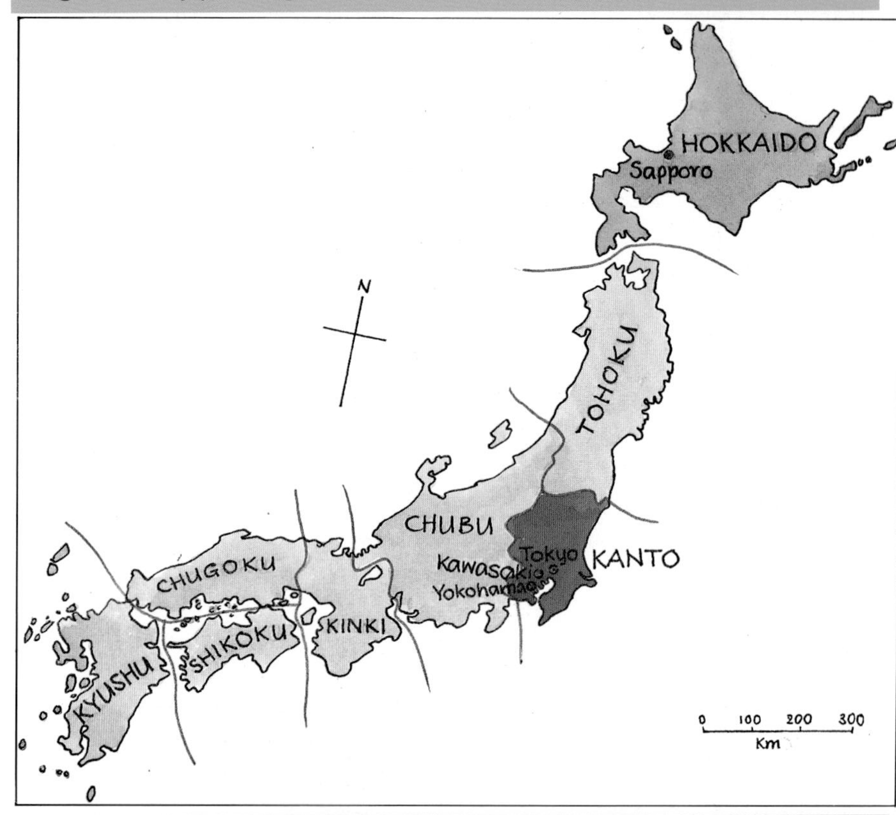

Figure 7.13 Comparing Hokkaido and Kanto: terrain, population and employment

Region	Area (km²)	Area covered by mountains (%)	Population (millions)	Population density (people per km²)	Income per person in 1988 (1000 yen)	Employment in primary industry (%)	Employment in secondary industry (%)	Employment in tertiary industry (%)
Hokkaido	83 500	49	5.6	72	2197	12.6	24.6	62.8
Kanto	32 000	40	38.6	1203	2703	7.5	35.5	57

Figure 7.14 Climate graph for Sapporo

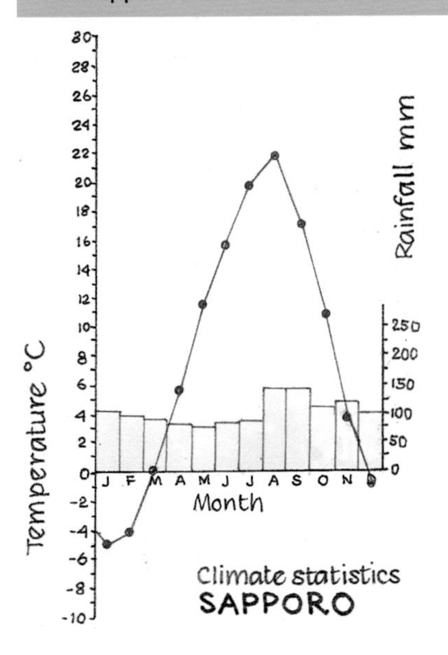

Farming is very important. Hokkaido's climate makes it the only region in Japan which is good for dairy and beef cattle, temperate cereals (e.g. wheat) and root crops (e.g. carrots). Other primary industries are coal mining, fishing and forestry.

Secondary industry is less important – the region accounts for less than 3 per cent of the country's manufacturing output. Most of its industries are to do with processing farm products, timber and fish, but there is also some shipbuilding, steel making and oil refining.

Tourism has become increasingly important. The region has superb scenery and excellent opportunities for winter sports. The 1972 Winter Olympic Games were held at Sapporo.

Hokkaido has benefited from **regional aid**, much of which has been spent on improving road, rail and air links. In 1988 the Seikan Tunnel was opened which connects Hokkaido with Honshu.

This has made the island much more **accessible** (a four hour ferry crossing has been replaced by a two hour rail crossing) and this is helping to attract new high-tech industries which Hokkaido needs if it is to make further progress.

Kanto

The Kanto region is made up of steep mountains to the north and west and the Kanto Plain, Japan's largest area of continuous lowland, to the south and east. Summers are hot, wet and humid while winters are cool and dry (Figure 7.15).

The availability of flat land helps to explain why Kanto is Japan's most densely populated region. It has three of the country's largest cities – Tokyo, Yokohama and Kawasaki – which have merged together to form a **conurbation** with a population of more than 12 million.

Kanto accounts for 36 per cent of the country's manufacturing output. It has a wide range of industries, such as oil refining, iron and steel, and electronics. It is also the country's financial and administrative centre.

There are so few areas of flat land in Japan that the Kanto Plain is important for farming as well as manufacturing. Fruit, vegetables, poultry, eggs, pigs and rice are the main products.

Unlike Hokkaido, Kanto's problem is too much growth, not too little. There is a lack of suitable land and it is overcrowded. There is **land use conflict** between, for example, farming, industry and housing.

The quality of life in Kanto is affected by pollution. In recent years, the government has passed strict anti-pollution laws and some progress has been made, e.g. levels of sulphur dioxide in the air have fallen. However, many problems remain, e.g. noise pollution is becoming worse.

Figure 7.15 Climate graph for Tokyo

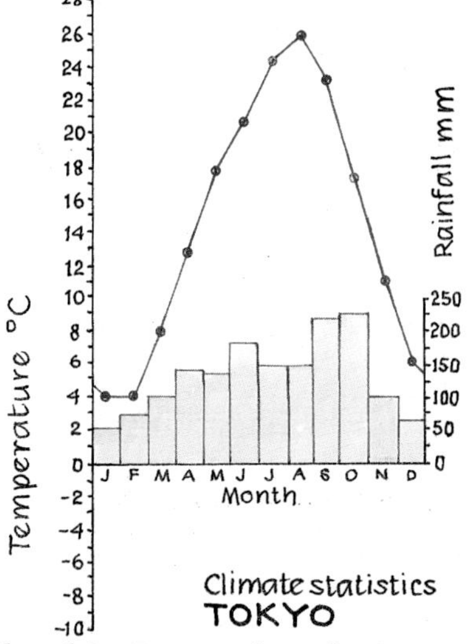

Word box

regional aid Help given by a government to one of its regions
accessible Easy to get to
conurbation A large urban settlement resulting from towns and cities spreading out and merging together
land use conflict Competition between different uses for the same piece of land

QUESTION BOX

1 On an outline map of Japan, mark and label Hokkaido and Kanto. Also label onto your map two ways in which these regions are similar and two ways in which they are different.

2 Write down one way in which Hokkaido is changing and one way in which Kanto is changing.

A Copy and complete Figure 7.16. Highlight similarities in red and differences in blue. Explain two of the similarities and two of the differences.

B If you were responsible for regional planning in a) Hokkaido and b) Kanto, what would be your three main priorities, and why?

Figure 7.16 Hokkaido and Kanto: a comparison to complete

	Relief	Maximum temperature	Minimum temperature	Total yearly rainfall	Population density	Income	Farming	Manufacturing industry	Tertiary industry	Main problems
Hokkaido										
Kanto										

Japan's economic links with the rest of the world

Since 1945 Japan has become one of the world's leading trading nations, for example, in 1991 it accounted for 9 per cent of the world's exports and 6 per cent of imports – these are remarkable figures when you think that it has only 2.4 per cent of the world's population!

Japan has few industrial raw materials, and very little coal and oil, so it had to import these when it started to rebuild its economy after the 1939–45 war. At the same time it began to export manufactured goods, especially textiles. Importing raw materials is still very important, although Figure 7.17 shows that the balance between the different types of import and export has changed since 1950.

In 1965 Japan had its first **trade surplus**; in other words, for the first time it had earned more from its exports than it had paid for its imports. It has enjoyed a trade surplus for most of the years since then: the figure in 1994 was $121 billion – countries can earn a lot of money from trade!

Figure 7.18 shows Japan's top ten trading partners in 1990. The USA has been its main trading partner since 1950 but its trade with Asian countries has increased to 50 per cent of the total.

In the 1980s Japan began to invest some of its wealth in other countries; for example, between 1986 and 1990 Japanese **transnational corporations (TNCs)** invested $227 billion, mainly in North America and Europe. The money was used to buy shares in foreign companies and set up factories (Figure 7.19).

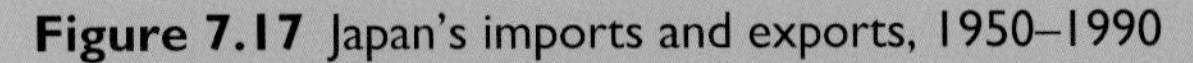
Figure 7.17 Japan's imports and exports, 1950–1990

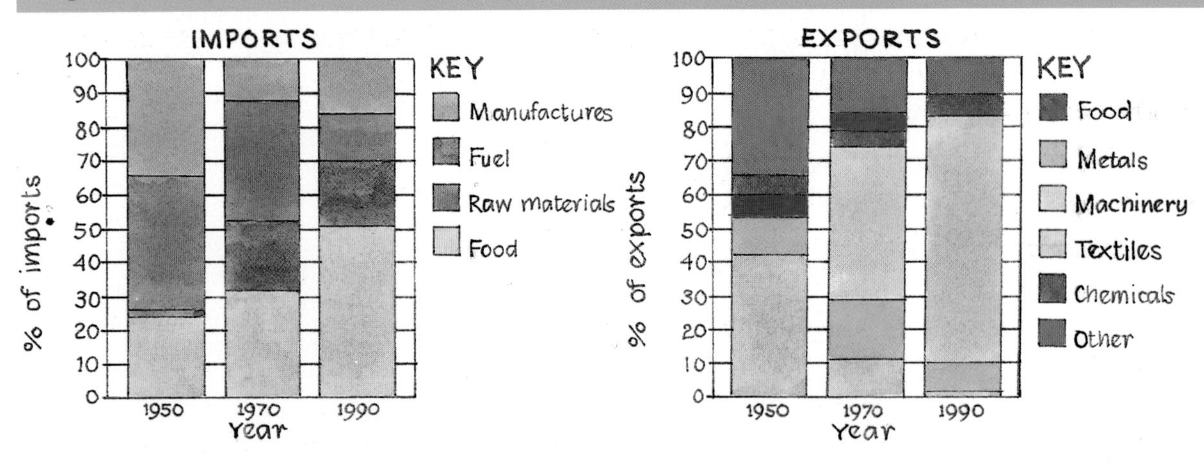

Figure 7.18 Japan's top ten trading partners, 1990

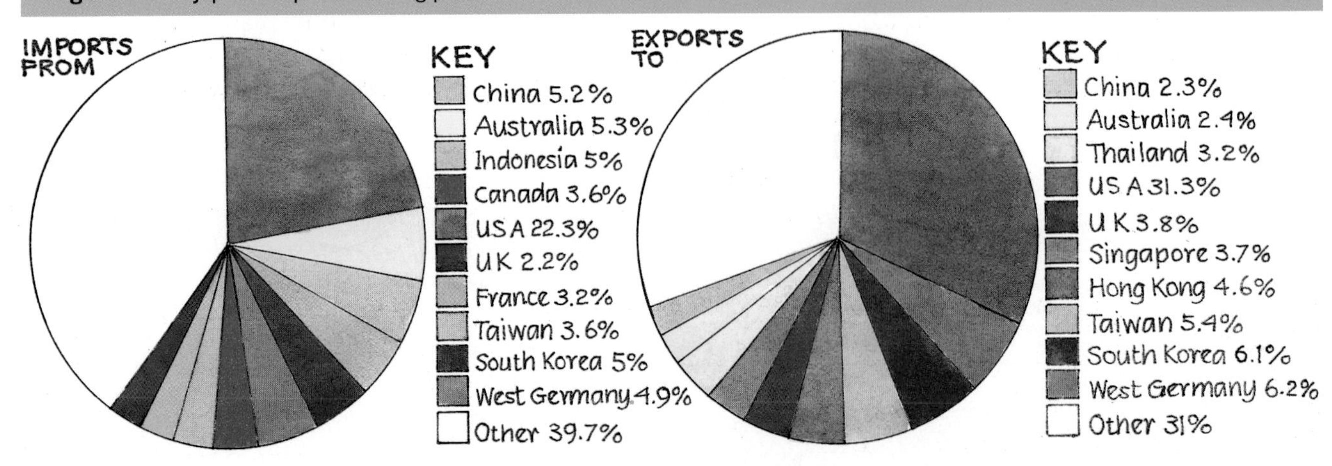

Figure 7.19 Japanese companies in Europe, 1992

Country	Number of Japanese Companies
Ireland	22
UK	120
Denmark	3
Netherlands	35
Belgium	23
West Germany	83
Luxembourg	3
France	87
Italy	28
Greece	3
Portugal	12
Spain	48

Figure 7.20 Honda car plant, Swindon

Japanese firms have benefited from overseas investment because it has opened up new markets for them. The countries in which they have set up have also benefited because the new factories have created jobs. However, there are disadvantages: e.g. employment in existing industries can be threatened, and much of the profit goes back to Japan.

The UK is a good example of a country which has received Japanese investment; in 1990 this amounted to US $6.84 billion. Most of this money was spent on buying shares in British businesses, e.g. Rover, but 15 per cent was spent on setting up manufacturing industries. Currently, Japanese factories in the UK employ 25 000 people. They have been attracted by good communications with the rest of the EU, the availability of **greenfield sites**, a skilled workforce, and government aid.

The first Japanese car plant in the UK was opened by Nissan in Sunderland in 1986 and was followed by others, e.g. Toyota on Deeside. Government grants have helped to attract electronics firms to South Wales, such as Sony at Bridgend.

Japan does not always get on with its business partners. The USA has wanted to close its **trade deficit** with Japan for some years and there have been serious disputes between the countries; e.g. in 1986 the USA put a 100 per cent **tariff** on $300 million worth of Japanese electronic goods.

Word box

trade surplus This is when exports are greater than imports

transnational corporations (TNCs) (or multinationals) Large companies with operations in more than one country

greenfield sites Sites which have not yet been built on

trade deficit This is when imports are greater than exports

tariff A tax which has to be paid to the government on an import

QUESTION BOX

1 What were Japan's two main imports and two main exports in a) 1950 and b) 1990 (Figure 7.17)?

2 How many Japanese companies were there in a) the UK and b) the rest of the EU in 1990 (Figure 7.19)?

3 Why have Japanese firms been attracted to the UK? Give one advantage and one disadvantage of Japanese firms setting up in the UK.

A How and why has Japan's pattern of trade changed since 1950?

B What advantages does Japan get from a) international trade and b) overseas investment? Why do these activities sometimes cause problems?

Change in Japan

Since 1945 Japan has changed a great deal. In the next 50 years it will develop in new directions as it deals with a range of issues, some of which are described in this section.

Social issues

- Many Japanese are concerned about westernisation which means the replacing of the traditional way of life with western customs, e.g. food, clothes and pop music. They are worried that Japan will lose the things which have made it so successful, such as hard work, discipline and respect for authority.
- Japan's high level of life expectancy combined with its low birth rate means that it has an ageing population – in other words, an increasingly greater percentage of its people are elderly (Figure 7.21). This means that a smaller **working population** has to support a larger **dependent population**.

Economic issues

- Japan is having to compete with the **Newly Industrialising Countries (NICs)** of Asia. Countries like South Korea have made manufacturing industry an important part of their development programmes and they have been very successful. If Japan is to stay ahead of them it will have to develop new high-tech products and almost certainly set up more factories in other countries.
- Tertiary industry has become more important in recent years, employing twice as many people as it did in 1970. Tokyo has become one of the world's main financial centres and it is likely that in the future Japan will earn an increasing amount of its money from activities like banking.

Environmental issues

- Manufacturing industry has brought with it the problem of pollution. Concern began in the 1950s when there were a number of serious incidents: for example, mercury poisoning from a fertiliser factory at Minamata affected over 2000 people. Japan now has some of the strictest anti-pollution laws in the world but there are still many problems to deal with.

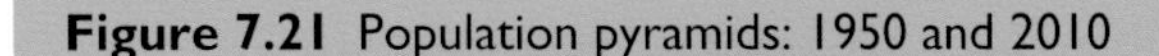

Figure 7.21 Population pyramids: 1950 and 2010

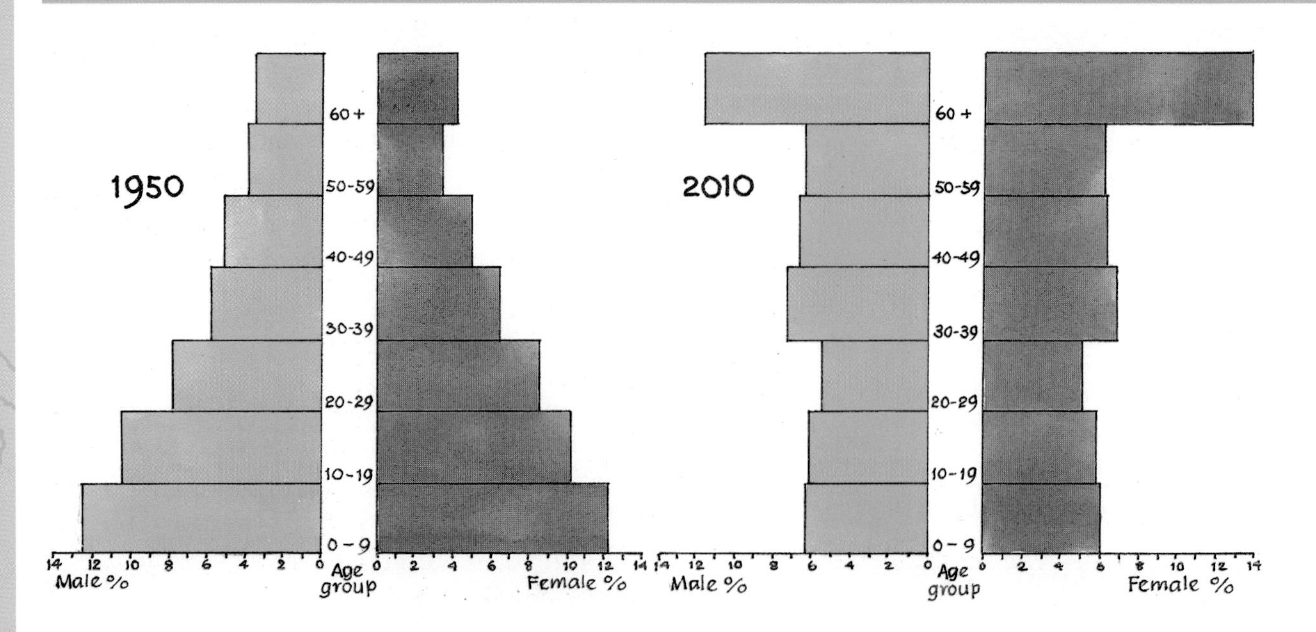

Figure 7.22 Mizushima industrial complex, near Kurashiki

Word box

working population People in employment

dependent population People who rely on the working population for support

Newly Industrialising Countries (NICs) LDCs which are developing their economies by setting up manufacturing industries

reclamation creating new land, e.g. from the sea, or cleaning up old land so that it is fit for use

- Land use conflict is a problem because there is so little flat land. **Reclamation** is a traditional solution: industrial areas, in particular, have been built on land reclaimed from the sea (Figure 7.22). Forests are being cleared, although they are supposed to be protected. For example, in 1988, 20 000 ha of forest was cleared, half of it for golf courses. Farmland is also being cleared – 50 000 ha in 1988, half of it for housing and industry.

QUESTION BOX

1 What is happening to Japan's population pyramid? Suggest reasons why this is a problem.

A What social and economic problems are being caused by Japan's changing population structure?

Extension task: see worksheet 7C/D

SUMMARY QUESTIONS

1 Make a list of the ways Japan has changed since 1945. Write a paragraph about two of these changes.

2 Carry out a survey of your house (or school!) and write down everything that was made by a Japanese firm. Use this evidence to help you to write about the type of products for which Japan has become well-known.

3 Find out, and write about, two of these traditional aspects of Japanese life: the tea ceremony; origami; martial arts; Noh theatre. Present your findings to the rest of the class.

A Why are Japan's links with the rest of the world so important to its economy?

B Which of the issues facing Japan (see pages 70–71) do you think is the most serious? How do you think it should be tackled?

C Find out about Japan's fishing industry. How large is the fleet? Where does it operate? What are some of the controversies about its activities? How is the industry changing? What are your views on the industry? Prepare a discussion paper on this topic for the rest of the class.

India's regions

Mountainous India

There are five mountain ranges which belong to the Himalayas in northern India. In the west there are 95 peaks that are over 7400 m high (Figure 8.5). The world's second highest mountain, K2, is in the Karakorams. The eastern Himalayas of Assam are lower and damper. This area has one of the world's highest rainfalls.

Figure 8.5 Foothills and snow-capped peaks of Himalayas

Figure 8.6 Vertical climates

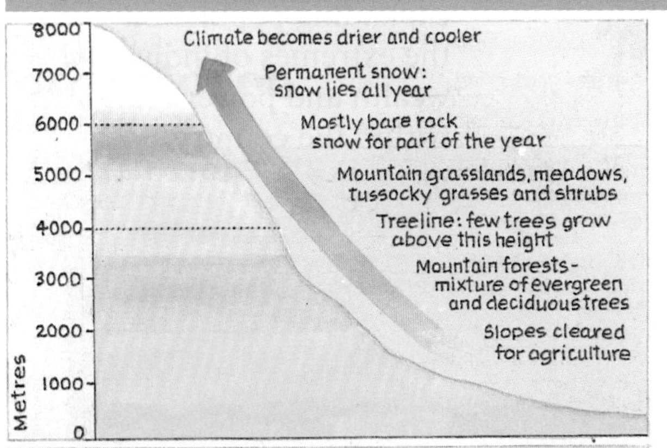

The mountains have a vertical range of climates, i.e. the temperature changes according to the height of the land (Figure 8.6). This area is sparsely populated. The land is farmed right up to the snowline and the environment is harsh and dry. Lower slopes have alpine meadows which are seasonally grazed by animals. In the foothills, cultivation of dry and wet **padi**, tea plantations, fruit trees and crops are found. The Himalayas have enormous potential for **hydro-electric power** (HEP) schemes. They are in an isolated area and remain unspoilt by human activity.

Figure 8.7 Ganges Plains

Figure 8.8 Case Study: Narmada Dam

Aim of Project: irrigation, hydro-electric scheme, water supply.

Project includes: 30 major dams, 135 medium dams, 3000 minor dams, 75 000 km of canal. The Narmada canal is 75 m wide and 439 km long, irrigating 1.8 million ha, supplying drinking water to 131 towns/cities, 4720 villages, supplying electricity to over 40 milion people in the region.

Cost: over £5 billion, employing 25 000 labourers, Due for completion in 1996.

Problems:
- Over a million people had to be moved as their land and villages were flooded.
- Environmental damage with the loss of habitat of rare plants and animals.
- Deforestation will cause soil erosion.
- The benefits are short-term as the dams silt up.
- The area is seismic and there is an increased threat of earthquakes as the lakes fill up.

Figure 8.9 Rajasthan and Kerala data

	Area (km^2)	Population (millions)	Population density	Literacy rate		GNP per person US $	Infant mortality per '000	Life expectancy (years)	Population growth rate (%) 1991–2001
				male %	female %				
Kerala	38 863	29	749	93	86	265	20	70	15
Rajasthan	342 239	44	129	54	20	290	80	58	27
India	3 288 000	903	278	64	39	303	88	60	24

The Plains

This low-lying, wide river plain extends eastwards across India from the Thar Desert to the Ganges Delta. The Rivers Ganges and Brahmaptura and their numerous tributaries criss-cross the plain (Figure 8.7). The rich **alluvial** deposits make fertile soils, and **irrigation** schemes, e.g. the Ganges–Kobadak scheme, help to increase agricultural production. The Plains form the most densely populated area of India. Wealth from farming encouraged industrial and urban development, e.g. The Domador Valley coalfield.

Rajasthan is one of India's least developed regions. It is a semi-desert area with no rain in the west and less than 60 cm in the east. The unpredictable rainfall makes farming difficult. The standard of living is low and in many areas farming is only possible with irrigation. Villages lack development and few have electricity. The region has a high birth rate and one of the lowest **literacy levels**. The main industries are mineral ore mining and cash crops. Ahmedbad is an important cotton and textile centre.

Peninsula India

South of the Ganges Plain is the Deccan Plateau, flanked by the Western and Eastern Ghats. These hills are forested. The volcanic soils are watered by the monsoon rains and this is a rich farming area. Water control schemes like the Narmada Project have provided hydro-electric power and irrigation Figure 8.8).

The southern part of India is tropical. Fishing villages with coconuts palms line the coast. The Kerala coast is an attractive holiday area. The region exports a variety of crops including exotic fruits, spices, sugar cane, rice, **coir** and millet. Twenty-five per cent of Kerala's income comes from money sent home by migrant workers in the Gulf. Kerala is a poor state in a poor country, but the population have achieved a higher **quality of life** than in the rest of India. Despite a high population density, Kerala has a low birth rate and longer life expectancy than the rest of India. The Kerala State Government has introduced a wide range of social reforms to benefit the population. These include land reform, with 90 per cent of the people owning the land their house is built on. Fair price shops enable people to buy basic food cheaply. The State spends 30 per cent of its budget on education. However, despite this Kerala has one of India's highest unemployment rates.

Word box

padi Rice
hydro-electric power (HEP) Electricity produced when running water from a reservoir or river is used to turn the blades of a turbine which is connected to a generator
alluvial River deposits
irrigation Adding water to land that is normally too dry to grow certain crops
literacy levels The number of people who are able to read and write
coir The fibre from coconut husks
quality of life A measure of how well people's basic needs are being met

QUESTION BOX

1. Use Figure 8.6 to describe how the natural vegetation changes as you go up the mountains.
2. Write down three facts about The Plains.
3. Describe what Kerala is like.

A. Explain why the natural vegetation changes as you rise.
B. Describe how The Plains landscape changes from the western to the eastern side.
C. Use Figure 8.8 to assess the likely benefits and problems of the Narmada project.
D. Use Figure 8.9 to compare the quality of life in Kerala with Rajasthan.

Extension task: see worksheet 8A/B

Rural and urban contrasts

India has been described as a land of villages. Over 650 000 villages exist with 70 per cent of the population still living in the countryside. Most are subsistence farmers or agricultural workers for landowners and 30 per cent of India's income comes from agriculture. Life in the countryside is difficult. Incomes and the standard of living are low and large numbers live in rural poverty. There is a range of climates within India but agriculture is dominated by the monsoon. There are five main **cropping zones** which are linked to the rainfall and landscape (Figure 8.11).

The farming production has been improved by irrigation schemes, particularly in the north west. Since India's independence, ground water has been used to irrigate crops (Figure 8.12). Crops yields have been increased by using better seeds, fertilisers and pesticides. Wheat production has increased from 6 million tonnes in 1947 to 55 million tonnes in 1993. These farming improvements are part of the **Green Revolution**. India is increasingly dependent on **intensification** of farming. New and **marginal land** is brought into production using irrigation and fertilisers. India's forested areas have been reduced from 17 per cent (1975) to 12 per cent (1990). The improvements in farming have not always benefited the Indian farmer or labourer (Figure 8.13).

Figure 8.10 The rice harvest in Sikkim, India

Figure 8.11 Rainfall and cropping zones

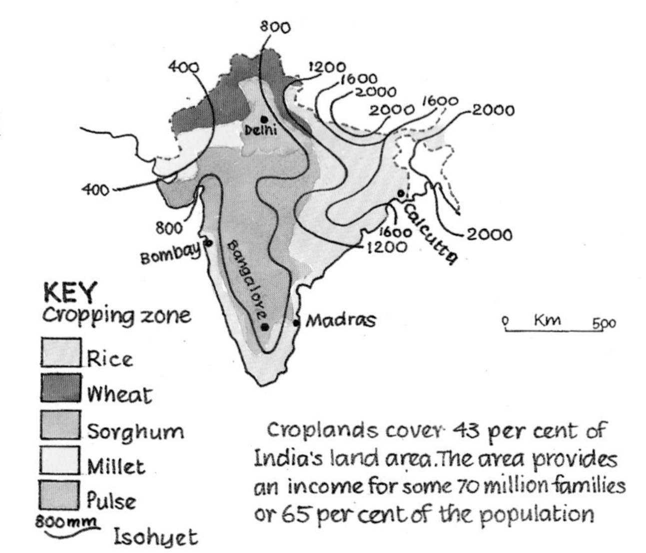

Figure 8.12 Tube well irrigation in west Punjab, India

Word box

cropping zone An area with one major crop

Green Revolution Improvements in farming by using better seeds, irrigation, pesticides, fertilisers and machines. These new ways increase crop yields

intensification Getting higher production from the same land

marginal land Poor farmland with infertile soils or harsh climate

Urban India

Seventeen per cent of India's population lives in cities. The urban population has increased by 80 million in the last ten years. In 1951, 62 million Indians lived in cities, this has increased to 238 million in 1991. The Green Revolution has reduced the number of farmworkers needed because of an increased use of machinery. The unemployed labourers are attracted to the cities hoping for work and a better life.

Figure 8.13 Winners and losers from the Green Revolution

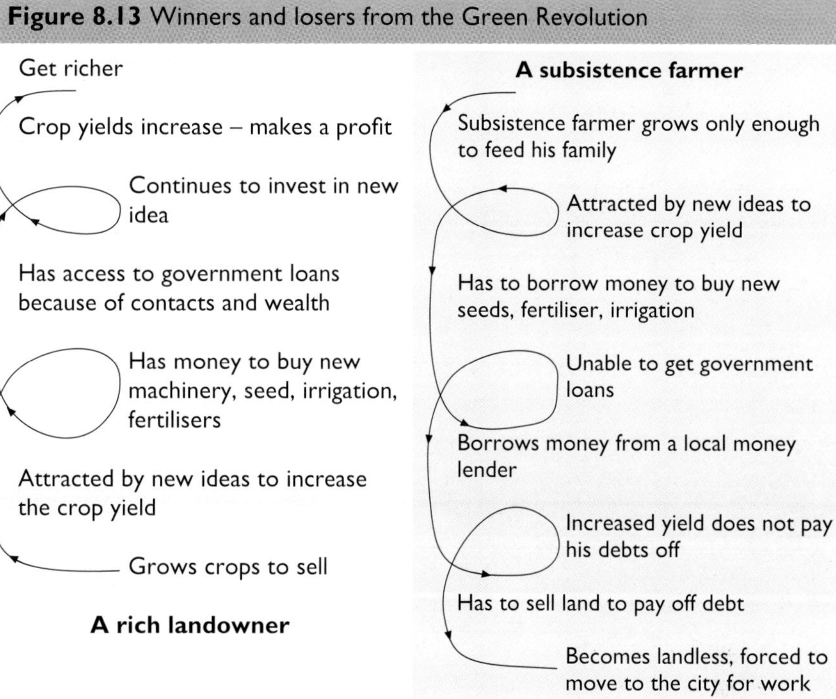

The migrants to the city find that conditions are often worse than in the countryside (Figure 8.15). Housing conditions are poor and jobs are low paid with high unemployment. Large areas of shanty towns grow up near industry and on the edge of the city. There are marked contrasts between and within urban areas. While a few live in luxury, the vast majority of people live in urban poverty. The infrastructure of Indian cities is unable to cope with the urban growth. For example, Calcutta's water and sewerage system was designed for a city of one million, today it copes with over nine million.

Some shanty towns are being improved with money from foreign aid. In many cities self-help schemes to rebuild housing and provide clean water and sanitation have improved some areas. However, the shanty towns are overcrowded and growing too rapidly for widespread improvements to take effect.

1 Look at Figure 8.10 and describe the appearance of the Indian village. How does the village differ from where you live?

2 What is the minimum amount of rainfall needed to grow rice?

3 Suggest why farming is important in India.

4 Suggest why people are leaving the countryside to live in the cities.

A Explain the causes of rural poverty in India.

B Use Figure 8.11 to describe the pattern of crops in India.

C Explain why subsistence farmers have not benefited from the Green Revolution in the same way as the rich landowners.

D Why have shanty towns developed in many Indian cities?

Extension task: see worksheet 8C/D

Figure 8.14 Urban plan of Delhi

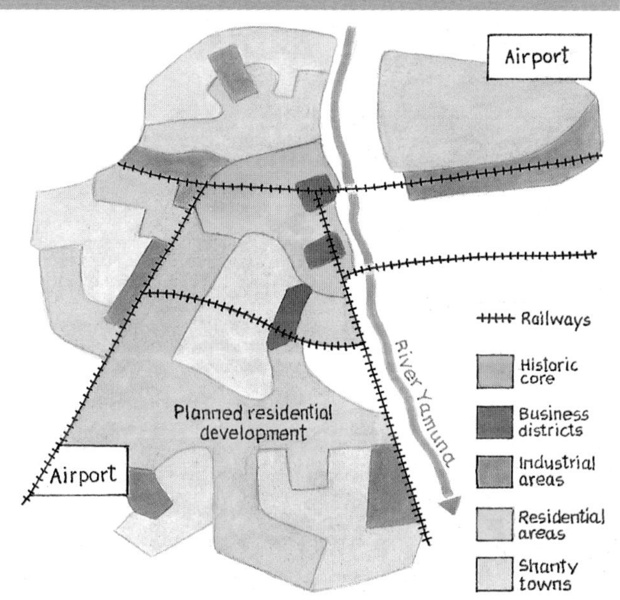

Figure 8.15 a) Shanty towns, Delhi b) Modern blocks near Connaught Place, Delhi

Industrial development

India at independence was a very poor nation. It had virtually no industry, it suffered from famines and poor health with a life expectancy of 28 years. These factors were obstacles to development. Since independence, India's industrial development has been achieved by a series of five year plans. India was supported by the former **eastern bloc communist states**. All decisions were taken by the government and major industries, e.g. steel production and cotton, were largely government controlled. Only small-scale industry and agriculture was owned privately. Foreign investment and ownership was limited as the Indian Government attempted to develop Indian industry.

India now has the twelfth highest GDP in the world and a modern skilled workforce. It has the capacity to make a wide range of products including high-tech consumer items. The increase in economic growth has classed India as a **Newly Industrialising Country** (NIC). Between 1970 and 1991, the contribution of agriculture to India's GDP fell from 45 per cent to 31 per cent, while industry has increased from 31 per cent to 45 per cent.

Despite rapid industrial growth, India is still one of the world's poorest nations. The increase in wealth benefits those employed in the industry and services found in the urban areas. The majority of Indians live in the rural areas and have seen no rise in their incomes.

India in the past has borrowed money from The World Bank and The International Monetary Fund (IMF) to pay for industrial development. India owes these banks over $800 billion. Since 1990, the Indian Government has changed its industrial policy

Word box

eastern bloc communist states The former communist USSR and eastern Europe countries

Newly Industrialising Country (NIC) Less developing countries which are improving their economies by setting up manufacturing industries

central planning Where a government makes all the decisions about the economic development in a country

transnational corporations (or multinationals) Large companies with operations in more than one country

Figure 8.16 Transnationals aim to persuade rural dwellers to adopt western products such as soap and shampoo

away from **central planning**. Foreign companies are being encouraged to invest in and work with Indian industry. The involvement of foreign **transnationals** in India will help reduce the need for more loans and an increasing debt. Many Indian business people are worried about the increasing influence of western and Japanese companies. Many fear that India's industry will suffer as Indians switch to foreign products like Coca-Cola (Figure 8.16).

India needs to work with foreign companies. They have the money to modernise India's industry and an example of this co-operation is the car industry. Obsolete technology and bad designs produced cars which had poor road performance and high fuel consumption. Suzuki worked with the Indian Government to produce the first all Indian car, the Maruti (Figure 8.17). The car was an instant success and since then, other transnational car companies have become involved in joint ventures. The opening of the Bombay Stock Exchange has been a sign of greater trade with the outside world.

Figure 8.17 The Maruti car

QUESTION BOX

1 List some reasons why India has found it difficult to develop industry.

2 Why is India called a Newly Industrialised Country?

3 Why did India have to borrow money from the IMF?

A Explain how India has tried to develop industry since independence.

B The encouraging success of the Maruti car shows what could be achieved. What are the benefits for India by working with a trans–national company? What are the problems for India by working with a trans–national company?

Changing India

A major problem faced by developing countries is how to provide the population with a good standard of living. India's population has grown rapidly and this has put pressure on resources and amenities (Figure 8.18). Despite the poverty, the death and infant mortality rates have improved and life expectancy has increased because of better health care.

Figure 8.18 India's population data

Census year	1901	1911	1921	1931	1941	1951	1961	1971	1981	1991	2001
Population in millions	235	249	248	275	314	359	438	547	688	827	980
Birth rate	49	46	46	43	44	41	42	39	36	31	25
Death rate	42	44	36	31	30	27	22	16	14	10	9
Infant mortality	250	—	—	—	—	190	183	145	90	88	69
Life expectancy	22	20	26	26	28	32	41	49	54	60	61
Percentage of the population living in Urban areas	11	10	11	12	14	17	18	20	28	34	27

A key issue in India's development is employment. India earns money by selling goods to other countries. This money is used with the tax income to pay for India's development. Many exports are low value agriculture products (Figure 8.19). The majority of Indians rely on agriculture for employment. This is a low paid occupation and it is becoming difficult to make more land available for farming. The Green Revolution means that fewer farmworkers are needed. Industrial development can only be achieved with the help of foreign investment. Modern factories use expensive and labour-saving machinery so only a small number of better paid jobs are created.

Figure 8.19 India's trading partners and some important imports and exports

Trading partners

Imports	%	Exports	%
Japan	13	USA	19
USA	10	CIS	15
Germany	9	Japan	10
United Kingdom	8	Germany	6
CIS	5	United Kingdom	6

Goods traded

Imports	%	Exports	%
Machinery	18	Gems and jewellery	17
Mineral fuels	13	Clothing	9
Iron and steel	7	Leather products	6
Pearls and precious stones	7	Machinery	6
Electrical machinery	4	Cotton fabric	5
Transport equipment	3	Tea	4
Edible vegetable oil	3		

QUESTION BOX

1. On a blank world outline, draw flow lines to show where India exports to, use a scale of 1 mm = 1% for the width of the line.
2. On a blank world outline, draw flow lines to show where India imports goods from, use a scale of 1 mm = 1% for the width of the line.

A Plot, as bar charts, India's imports and exports. Describe the nature of India's trade with the rest of the world.

B Explain why unemployment may increase as India becomes more industrialised.

Word box

informal employment Unofficial, irregular work

Development Plan The Indian Government's five year industrial development strategy

self-help scheme Where local people are provided with resources and help to improve an aspect of their lives or livelihood

intermediate technology A simple, low-cost improvement to a tool and or way of making something which uses local materials and is done in a way which ordinary people understand

Many people do not have a regular paid job. Instead they find casual employment as street traders or by providing a service, for example, rubbish collecting and sorting. This **informal employment** is a very important source of income.

India's previous five year plans favoured large-scale industrial complexes such as the Damodar coalfield centred on Asansol, steel making at Jamshedpur and Durgapur, oil refining at Barauni, cars and trucks in Madras and Calcutta, and heavy electrical engineering in Hyderabad and Bangalore. Industrial development has been urban based. Several nuclear power stations are in operation and the enormous potential of hydro-electrical power is being exploited. Many towns and cities have trading estates with nearby residential areas.

Whether large-scale industrial development or schemes like that of the Narmada Valley are the best way for India to provide employment is being questioned. The Fifth **Development Plan** of 1974–79, placed greater emphasis on rural craft industries. These small-scale schemes may prove to be the best way to make rural areas more prosperous. Such schemes would employ the large rural workforce, many skilled in traditional crafts, and use a variety of natural resources.

Cotton is one of India's most important exports. Calcutta and Ahmadabad are large cotton centres, although many villages produce a wide range of cotton and other craft products. These crafts can provide alternative employment in farming villages and encourage people to stay instead of migrating to the cities With the help of aid organisations and overseas business people, many villages have organised themselves into **self-help schemes** and co-operatives. Their use of **intermediate technology** helps in the production but does not replace people.

SUMMARY QUESTIONS

1. Suggest why India could be described as a land of contrasts.
2. Why can India be described as an LEDC?
3. What is meant by the Green Revolution?
4. Suggest why India remains one of the world's poorest nations.
5. Suggest why India's rapid population growth will be a threat to its economic progress.
6. Why are rural areas being encouraged to develop craft based industries?

A. Use the information to produce an annotated map of India's physical regions. Use an atlas to add further information about places and climate.

B. Explain how you would persuade an Indian subsistence farmer to invest in new ideas to increase his crop yield.

C. What are the disadvantages of basing industrial developments only in urban areas?

D. Use Figure 8.18:

a) plot the birth and death rates as a line graph

b) plot India's population as a line graph

c) plot life expectancy as a bar chart

d) describe the population trends for India.

E. Explain three problems which the Indian Government must overcome to ensure the benefits of a higher level of industrialisation for India.

CASE STUDY

Fiat and the Italian car industry

Fiat dominates Italy's car industry. It was founded in Turin (Figure 9.9) in 1899 and it is now one of the world's biggest industrial groups. In Italy, it employs 133 000 people and it makes three quarters of all the country's vehicles. It owns, or is involved with, nearly all of the country's other car manufacturers, like Alfa Romeo and Lancia. Sixty three per cent of its 1998 earnings of 89 billion lire came from export sales.

Figure 9.9 Fiat car works in Turin, Italy

Figure 9.10 Fiat factories in Italy

Until the 1980s production was concentrated in the north of the country (Figure 9.10). The advantages of this region included flat land for building factories, good roads and railways, and a large population/workforce. Fiat's headquarters is in Turin and 60 per cent of this city's jobs still depend on the company.

However, in the 1980s Fiat began to set up factories in other parts of the country, including the poorer south, which is known as the **Mezzogiorno** (see pages 94–5). They have done this for a number of reasons: the industrial cities of the north have become increasingly congested and polluted; *autostrada* (motorways) have made the south more **accessible**; there is plenty of flat land available for expansion; and there has been help from the government and the European Regional Development Fund in the form of grants, loans and tax concessions.

Many of Fiat's factories make **components** which are sent to other factories for final assembly. This helps to explain why good communications have become important to the location of the car industry in recent years.

Fiat is also one of the top 20 **TNCs** in the world. One hundred of its 185 factories are outside Italy and worldwide it employs 221 000 people. Almost 40 per cent of its output is manufactured outside Italy.

Fiat began to set up in other countries in the 1950s when it invested in Spain, and then in other parts of Europe and North and South America. However, it is only in the last ten years that Fiat has really taken off as an international company. In particular, it has invested in a number of LEDCs, e.g. Brazil, Argentina, India and China.

Fiat's 'new world car', the Palio, is an important part of its plans for overseas expansion. Production began in Brazil in 1996 and it has plans to produce one million Palios a year worldwide.

Figure 9.11 Fiat's Worldwide Presence

Factories		Employees	
Italy	85	Italy	132 688
EU (excluding Italy)	35	EU (excluding Italy)	31 655
Other European countries	13	Other European countries	15 459
Mercosur*	22	Mercosur*	30 940
NAFTA**	16	NAFTA**	7376
Other regions	14	Other regions	2421
Total	185	Total	220 549

*Mercosur (the Common Market of the South) = Brazil, Argentina, Uruguay, Paraguay
**NAFTA (North American Free Trade Agreement) = Canada, USA, Mexico

QUESTION BOX

1. Look at Figure 9.10. Describe the distribution of Fiat factories in Italy in a) the 1970s and b) the 1990s.
2. How has the distribution of Fiat factories changed? Suggest reasons for this change.
3. Write down three reasons why Fiat can be called a **TNC**.

A Describe and explain the changing distribution of Fiat's factories in Italy since 1970.

B Select and use appropriate techniques to present the statistics in Figure 9.11.

C What do these statistics tell us about Fiat as a **TNC**?

Word box

TNCs Transnational Corporations: Large companies with operations in more than one country

Mezzogiorno 'Land of the midday sun'. The poorer, less economically developed south of Italy

accessible Easy to get to

components The parts which make up the whole, e.g. engine, gear box etc. for a motor vehicle

CASE STUDY

The Brazilian car industry

The car industry has played an important part in Brazil's economic development. The first factory was built by Ford in 1919 in São Paulo and by 1924 it was producing 24 000 Model 'T's a year. However, these cars were only assembled (put together) in Brazil – all of the components were imported from the USA.

In the late 1950s the government decided to develop manufacturing industry in order to cut imports. It had neither the money nor the skills to set up large numbers of factories itself so it attracted **TNCs** with grants, loans and low taxes. At the same time, it discouraged imports with high **tariffs**.

Now, the car industry produces 1.68 million passenger vehicles a year (1997) which makes it South America's top, and the world's eighth largest, vehicle manufacturer – it really has become an important part of the Brazilian economy!

As far as its location is concerned, it is still concentrated in the São Paulo district where the industry began in the 1920s (Figure 9.12). For example, Ford has spent $1.1 billion since 1995 to build the Fiesta at its São Bernardo factory near São Paulo.

However, there has been some **decentralisation** in recent years. For example. Fiat opened a factory in Belo Horizonte in 1978 and more recently Renault has built a $1 billion dollar factory at Curitaba.

Figure 9.12 The location of the car industry in Brazil

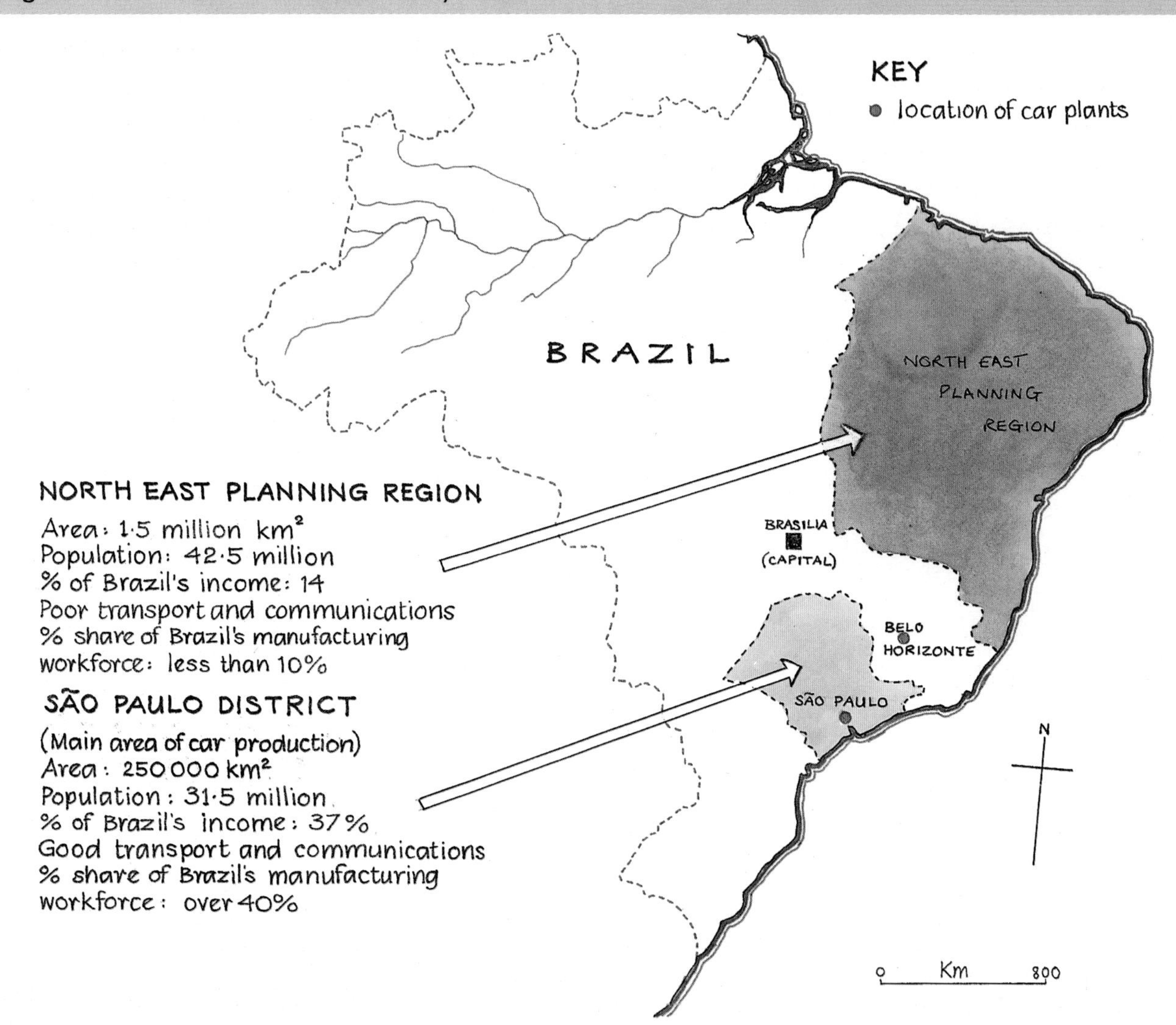

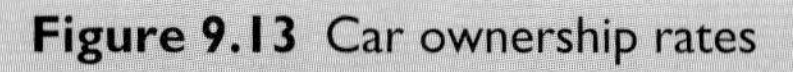

Figure 9.13 Car ownership rates

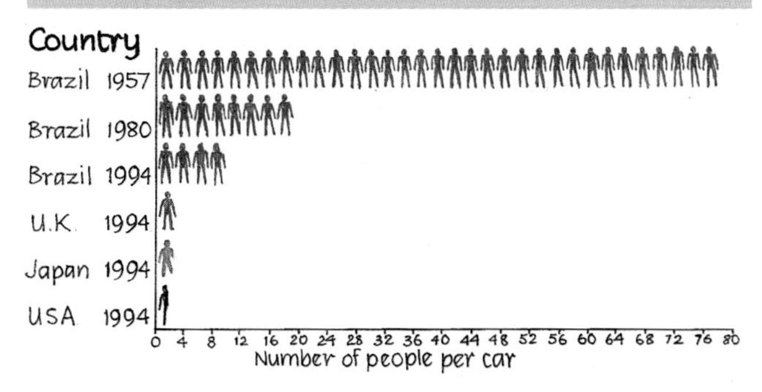

Also, Brazil is still dominated by **TNCs**. The major producers are General Motors (from the USA), Volkswagen (from Germany), Ford (from the USA) and Fiat (from Italy). They have created jobs, brought modern technology to the country and cut imports, but a lot of the profit goes back to their headquarters overseas and they have been criticised for paying very low wages.

Mercosur (the Common Market of the South) has helped Brazil's car industry to develop. It is a **trade bloc** which was set up by Brazil and Argentina in 1987. Uruguay and Paraguay have become members since then. The countries have reached agreements which have made it easier to trade between the countries. It has meant that **TNCs**, such as Fiat, can set up in just one of these countries but sell cars to all of the others without having to pay **tariffs**, or be restricted by **quotas**. One of the reasons why Renault set up in Curitiba was because it is easy, with Mercosur agreements in place, for it to get parts and workers from the factory it part–owns in Cordoba in Argentina.

The future for Brazil's car industry looks promising. Although it is a poor country, (see pages 96–7), average incomes are going up and each year more people can afford to own a car (Figure 9.13). The big car companies are confident that demand will continue to rise and many of them are planning new plants, e.g. Honda, Toyota, Chrysler, Hyundai and Kia (Figure 9.14).

Figure 9.14 Western car factory in Brazil

QUESTION BOX

1. Where is Brazil's car industry mainly located? Suggest reasons for your answer.
2. Give two reasons why Brazil's car industry has grown in recent years.

A To what extent has the distribution of Brazil's car industry changed since 1950? Give reasons for your answer.

B What is Mercosur and how has it helped Brazil's car industry to develop?

Extension task: see worksheet 9C/D

Word box

tariff A tax which has to be paid to the government on an import

decentralisation A move away from an existing centre

quota A limit on the amount of something which can be imported

TNCs Transnational corporations: large companies with operations in more than one country

trade bloc countries which group together for the purposes of trade

The world pattern of trade

Figure 9.15 The world pattern of trade

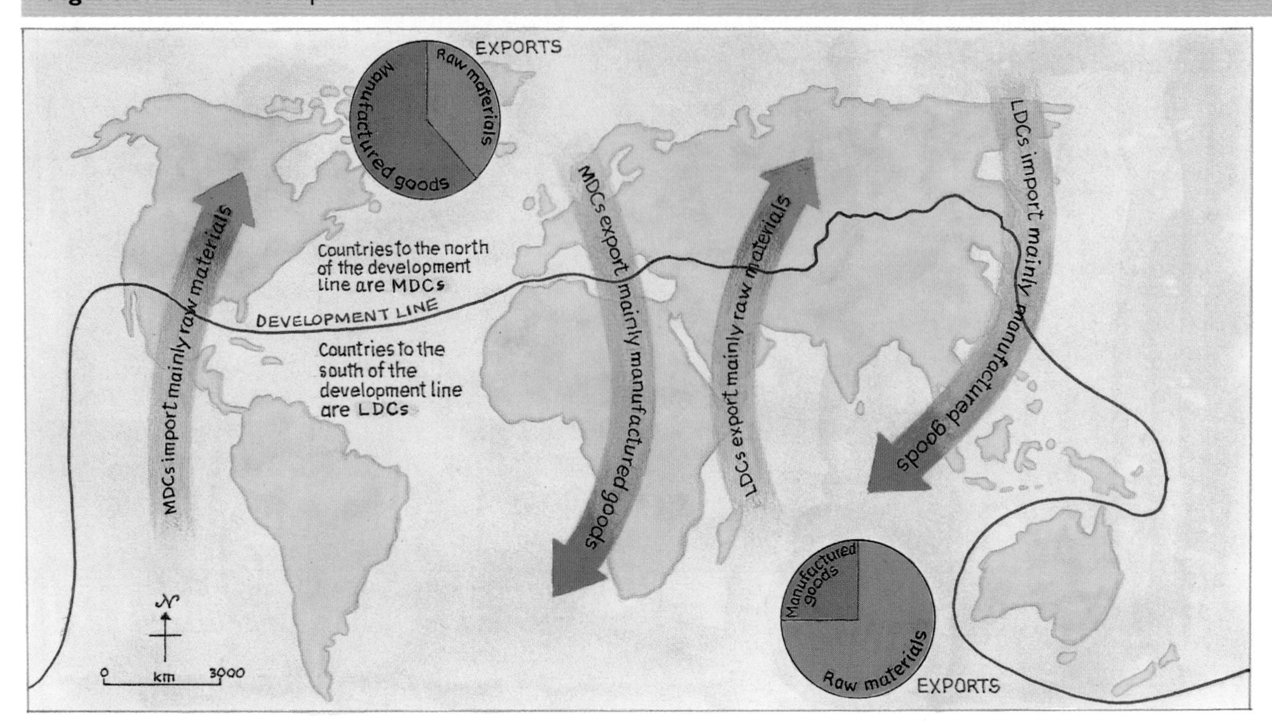

Countries trade with each other for a number of reasons: they do not have everything that they need or want; it is a way of earning money; and it is a way of building good international relationships.

World trade became important in the nineteenth century when Europe and North America began to import food and raw materials from South America, Africa and Asia; and in return to export manufactured goods. To a large extent, this pattern of trade still exists (Figure 9.15). It is unfair on LEDCs who find themselves in a 'trade trap' – a situation where manufactured goods cost more than raw materials and so their imports cost more than their exports. This makes it difficult for them to buy the machinery they need to set up their own manufacturing industries. However, LEDCs have been slowly increasing their share of world trade in manufactured goods as **Newly Industrialising**

Countries (NICs), have developed their own industries (see pages 84–85).

As world trade has become more complicated, so have the measures taken by countries to increase trade, or to protect themselves from foreign competition. **Trade blocs**, like **OPEC** or Mercosur (see pages 88–89) have become common. There are also international agreements, like the General Agreement on Tariffs and Trade (GATT), but these agreements can be hard to reach – the last round of GATT talks lasted seven years!

Word box

NICs (Newly Industrialising Countries) LEDCs which are developing their economies by setting up manufacturing industries

trade bloc Countries which group together for the purposes of trade

OPEC Organisation of Petroleum Exporting Countries: 13 oil producing countries in South America, Africa and Asia

interdependence Countries relying on each other, e.g. for certain products or services

MEDCs still dominate world trade – the value of their manufactured exports is seven times greater than that of LEDCs – but the degree of **interdependence** between MEDCs and LEDCs is increasing. For example, MEDCs are importing more and more components as well as finished products.

QUESTION BOX

1 Look at Figure 9.15. Describe the world pattern of trade.

2 Why is this world pattern of trade unfair on LEDCs?

A How, and why, is the pattern of trade shown in Figure 9.15 changing? (Use your knowledge of TNCs, **NICs** and the car industry in Brazil (including **trade blocs**) to help you with this answer.)

B Explain what is meant by 'economic interdependence'.

Extension task: see worksheet 9E/F

SUMMARY QUESTIONS

1 a) Find out which countries are members of the European Union (EU). Label them onto a suitable base map. Mark on and label Strasbourg (European Parliament) and Brussels (headquarters of the European Commission).

b) Find out about, and write down, three ways in which being a member of the EU has made trade easier.

c) Finally, find out the names of the MEPs (Members of the European Parliament) who represent your region.

A Find out more about **OPEC**. When was it formed? Why? Who are its members? What did it do in the early 1970s, and why? What effect did OPEC's actions have? How important is OPEC now?

unit 10

PROGRESS: BUT HOW?

Key questions

- Is economic development evenly spread at global and national levels?
- How can levels of development be measured?
- What causes underdevelopment?
- How useful are prestige projects?
- What are the advantages and disadvantages of investing in an LEDC?
- Why could intermediate technology benefit more people in an LEDC?

World development patterns

In 1980 the **United Nations** produced the Brandt Report which highlighted a wealth gap in the world between the richer, industrialised northern nations and the less industrialised and poorer southern ones (Figure 10.1).

The level of economic development and **quality of life** in a country can be measured by using a range of **indicators**. These include information about a country's population, wealth and economic features. Gross Domestic Product (GDP) is a useful measure of a country's development. It measures the value of all the work done in each sector of the economy. Gross Domestic Product per person (per capita) is the average value that each person has in the economy. Occupation structures with a high percentage employed in primary industries have low GDPs. Higher GDPs are achieved when a country has more people employed in manufacturing industries and services. This provides wealth to raise the country's quality of life. The country is able to provide healthcare, education and improved living conditions.

Figure 10.1 The North South divide

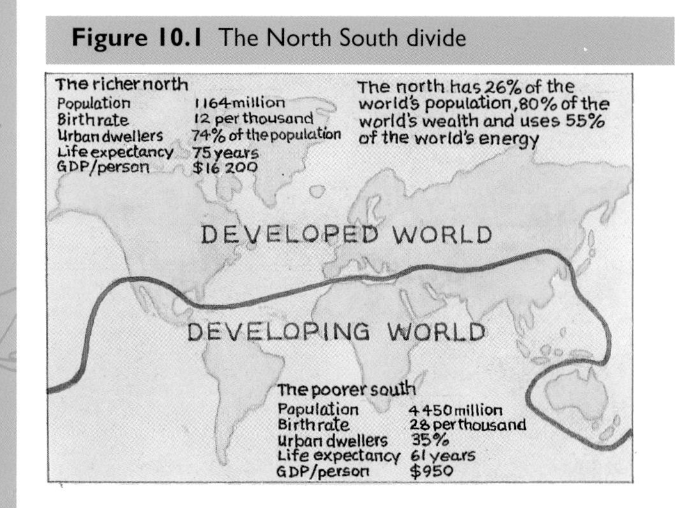

Reasons for the wealth gap

- *Physical setting.* Many poor countries have difficult environments. Droughts, floods, earthquakes and other **natural disasters** can ruin crops. The climate can be too hot or dry or have unreliable rainfall which makes successful farming a problem. Difficult relief may make parts of the country inaccessible. This prevents the exploitation of raw materials and the development of industry.
- *Historical background.* In the nineteenth century, many European countries colonised much of Africa, Asia and South America. The Europeans were looking for a cheap and reliable source of raw materials to supply their industries and these **colonies** became a protected market for their manufactured goods. The colonies were developed for the benefit of the Europeans but most colonies are now independent. The trading links with the

European countries are still strong, and European and North American companies still control many of the industries in the former colonies and especially the trade in raw materials.

- *Economic structure of developing countries.* Developing countries depend mainly on the primary sector to earn money and provide employment. Agricultural products and unrefined **raw materials**, e.g. iron ore and timber are exported to the industrialised nations. This type of export has a low value and does not provide sufficient capital for an LDC to develop the necessary **infrastructure** and industrial base to increase its export earnings.
- *Population.* As health conditions improve in developing countries, a rapid growth in the population occurs. This increase uses up resources which could have been set aside for economic progress. An increasing population needs money spent on health care, housing, jobs and other services to maintain a basic standard of living.

Figure 10.2 a) Peasant girl in India
b) Wealthy Gujarati business man, India

Word box

United Nations A world organisation which helps to maintain peace and improve the standard of living for the world's poorest peoples and nations

quality of life How well the population's basic needs, food, water, health care and shelter are met

indicators A way of measuring something about a country's social and economic state

natural disaster A violent natural event which threatens people

colonies Lands occupied and ruled by another country

raw materials Unprocessed and unrefined products such as iron ore, bauxite and rubber

infrastructure A range of communication networks providing services in an area, e.g. road and rail links, electricity and water supplies

QUESTION BOX

1. Identify the continents in the developing world.
2. State three differences between the developed and developing world.
3. Suggest some reasons why some countries are underdeveloped.

A. Explain what is meant by the term 'north–south divide'.

B. Explain how the economic structure of a country can affect its level of development.

Development in southern Italy

Regional differences

To many, Italy appears as a rich nation. It is home to many famous brand names and products. It is a world fashion centre and Italian history and art have influenced many countries. Italy has 20 regions which vary in character (Figures 10.3 and 10.4). There is a contrast between the richer north and poor south or **Mezzogiorno**.

The north developed as the main industrial region between 1950 and 1960. The main industries are located in and around a triangle of Turin, Milan and Genoa. Commercial agriculture became more efficient and modernised. The north attracted four million workers between 1951 and 1971. The lack of planning meant that the influx of migrants caused considerable urban problems, densities and run down areas.

Figure 10.3 Italy's North South divide

Lombardy
Population 8.9 million
% employed
in farming 4
in industry 47
in services 49
% of Italy's national output 27

The North
80% of industry
55% of population
80% of GNP

4 million moved north between 1951-1971
100 000 moved north between 1971-1981

KEY
Average incomes
Above national average
Below national average

Basilicata
Population 0.5 million
% employed in farming 30
in industry 25
in services 45
% of national output 0.5
% of population below poverty level 32

TURIN, MILAN, LOMBARDY, VENICE, BOLOGNA, ANCONA, ROME, NAPLES, BARI
0 Km 200

Figure 10.5 Italian regional indicators

Area key on map (Figure 10.4)	Region	% of workforce employed in agriculture	% of workforce employed in manufacturing	% of workforce unemployed	energy use kg/coal equivalent	number of persons per car	birth rate per 1000 people	Illiteracy rate % of the population	% of Italy's national output
1	Piedmont	10	43	5	9990	2.3	8	1.0	13.6
2	Valle d'Aosta	11	30	2	527	2.0	9	0.7	0.2
3	Lombardy	4	47	8	22 812	2.4	9	0.6	27.9
4	Trentino-Alto Adige	15	28	3	2100	2.7	11	0.3	1.5
5	Veneto	9	40	4	9 870	2.6	9	1.0	9.4
6	Friuli-Venezia-Guilia	8	33	5	2723	2.5	7	0.6	2.7
7	Liguria	8	28	8	2291	2.6	7	1.0	3.0
8	Emilia-Romagna	13	37	4	6774	2.2	7	1.4	9.1
9	Tuscany	7	39	7	6255	2.3	6	2.1	8.0
10	Umbria	11	37	8	1446	2.4	8	3.0	1.4
11	Marche	14	40	7	2266	2.5	9	2.5	2.6
12	Lazio	7	22	11	3282	2.5	9	2.0	5.3
13	Abruzzo	19	28	14	1589	3	10	4.6	1.4
14	Molise	28	25	14	316	3.7	11	5.5	0.2
15	Campania	16	27	22	3544	4.1	15	5.1	4.4
16	Puglia	22	25	14	5951	3.9	13	5.2	3.4
17	Basilicata	29	25	20	489	4	14	8.2	0.5
18	Calabria	22	22	19	1226	4.8	14	8.7	1.0
19	Sicily	19	23	24	5494	3.3	13	5.7	3.1
20	Sardinia	16	25	19	4466	3.5	10	4.6	1.3
	Italian average	12	34	10.8	4670	2.9	11	3.2	–

NB: *All figures have been rounded up*

Difficulties

Historically, the Mezzogiorno in the south has been a poor region. Before **Italian unification**, the south lagged behind, with many large farming estates exploited by foreign or northern landlords. The north was favoured for industrial development and government infrastructure schemes were built in the northern regions. The south lacked the necessary road and rail links to connect it to the rest of mainland Europe.

The Mezzogiorno is a mountainous region and lacks good farmland. Farming is difficult and summer droughts are common. Many farmers lacked **land rights** and worked on inefficient estates. Incomes were low and there were few opportunities to get paid work in a factory (Figure 10.5).

Help to the south

In 1950, the Italian Government set up a Fund For The South, *Cassa pes il Mezzogiorno*. Large sums of money were used to build roads and large industries, e.g. steel and petrochemical works. A number of **development poles** were planned. It was hoped that these industries would help create other jobs in the area. **Land reforms** changed the large estates into 120 000 new farms for local people. The government and the European Union subsidised the farmers and gave grants for irrigation and land improvement schemes.

Figure 10.5 Impressions of Mezzogiorno

Today

There has been an improvement in the standard of living in many parts of the south, for example in Toranto and Pescara. But there is still an imbalance between the south and the north, and the government has now disbanded the Fund For The South. To create jobs, small-scale industries, tourism and direct help for farmers has been encouraged. However, the **Common Agricultural Policy** (CAP) has benefited the larger northern, commercial farmers rather than the smaller family farms in the south. Progress in Sicily and Calabria has been further hampered by the influence of the **Mafia**.

The Mezzogiorno is gaining population from workers who return from the north or overseas for social and family reasons. The small family farms act like a sponge absorbing the returning workers and releasing them to paid jobs in the north when needed. Today Italy's poorly paid manual jobs are also being taken by immigrants from Asia, Africa and eastern Europe, adding to the unemployment problem in Italy.

Word box

Mezzogiorno 'Land of the midday sun.' The poorer, less economically developed south of Italy

Italian unification Before 1860 Italy was a number of states which then joined together to form Italy

land rights When people have a legal right to occupy or own farmland

development pole Industry which has been set up with the hope of attracting other industry to the area

land reforms A change in the way the land is owned and organised

The Common Agricultural Policy (CAP) A European Union scheme to help farmers and ensure food supplies are provided

Mafia A criminal organisation which has exploited and corrupted sections of Italian society

QUESTION BOX

1. Does Italy have a rich north and a poor south? Give a reason for your answer.
2. Describe what people think of the Mezzogiorno.
3. Describe how the Cassa pes il Mezzogiorno tried to improve the south.

A Select and present information from Figure 10.4 to demonstrate that a north–south divide exists in Italy.

B Assess how successful the schemes have been to improve the Mezzogiorno.

Patterns of wealth in Brazil

Figure 10.6 Brazil's major cities and industrial regions

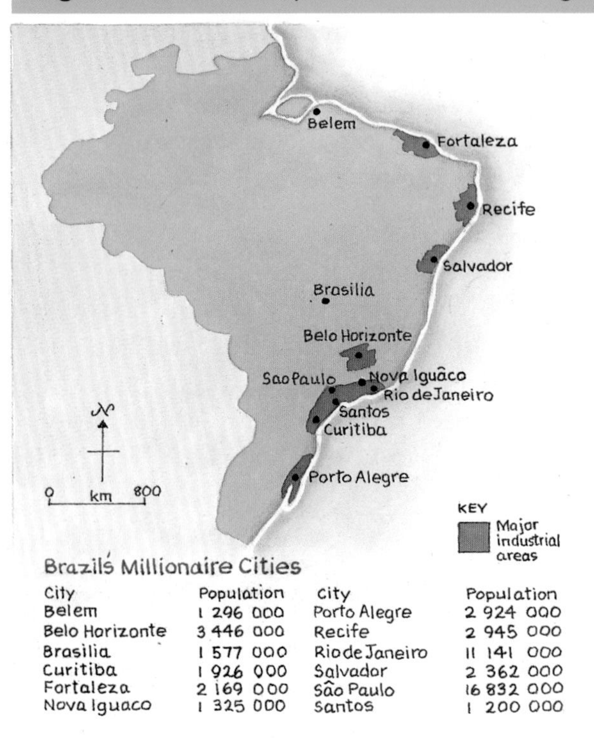

Brazil's Millionaire Cities

City	Population	City	Population
Belem	1 296 000	Porto Alegre	2 924 000
Belo Horizonte	3 446 000	Recife	2 945 000
Brasilia	1 577 000	Rio de Janeiro	11 141 000
Curitiba	1 926 000	Salvador	2 362 000
Fortaleza	2 169 000	São Paulo	16 832 000
Nova Iguaco	1 325 000	Santos	1 200 000

- Brazil is the world's fifth largest country with a population of 146 million. Despite its size, it has a population density of only 17 people per km^2 and there are large areas of land which are uninhabited and undeveloped. Seventy-five per cent of the population live in urban areas with 12 cities having over one million people (Figure 10.6). Brazil is a country of economic contrasts with both high-tech industries and very traditional methods (Figure 10.7).

Word box

core periphery One region, the core, is economically more developed than the rest of the country, the periphery, and the level of development diminishes with distance from the core

plantations Large commercial farms growing cash crops, e.g. bananas or coffee

vicious cycle of poverty Where people are trapped by being unable to earn enough to improve their standard of living

Regional wealth

Brazil is divided into five administrative regions. Economic development between the regions is very uneven. There is an economically developed **core** and **periphery** in Brazil. The south east region is the economic core of Brazil. It has the largest concentration of people and manufacturing industry. Greater São Paulo, for example, has 30 per cent of Brazilian industry and 45 per cent of manufacturing employment. The region also has above national average income. Away from the south east there is less development and the north and north east are the least developed (Figure 10.8).

Figure 10.7 a) Amazonian Indian village, Brazil

Figure 10.7 b) São Paulo city view, Brazil

Figure 10.8 Brazil's economic regions

	North	North east	Centre west	South east	South
% of population in area	5	29	6	44	16
% density per km^2	2.2	28.1	6	70.9	44
% of national income	2	14	3	64	17
% of industrial employment	1	10	2	70	17
% of cultivated land	1	18	25	33	22
Area, '000s of km^2	3581	1549	1879	925	578

Brazil is the world's second largest exporter of agricultural products. Seventy per cent of Brazil's farmland is owned by only 5 per cent of the population. The best land is found in the south and south east where there are large commercial **plantations** and farms. In the north east and Amazonia there are family subsistence farmers and conditions are very difficult. Agricultural workers have few services available or land rights. Wages are low. Many people are forced to leave the rural areas to live and find work in the industrial cities.

Many Brazilians are unemployed, underemployed or work as unpaid family helpers. They are unable to share in Brazil's wealth and are caught in a **vicious cycle of poverty** (Figure 10.10). With so many people looking for work, the wages stay low and their working and living conditions poor.

Figure 10.9 Simple core periphery view of Brazil

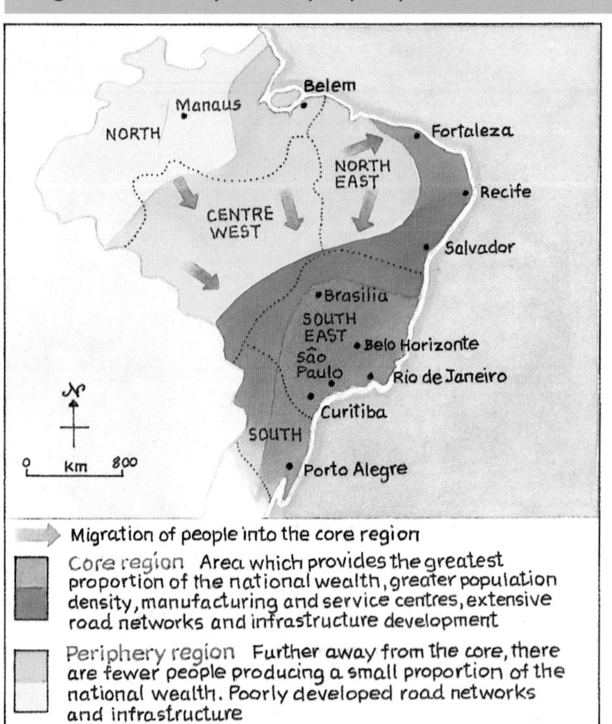

Figure 10.10 Cycle of rural poverty

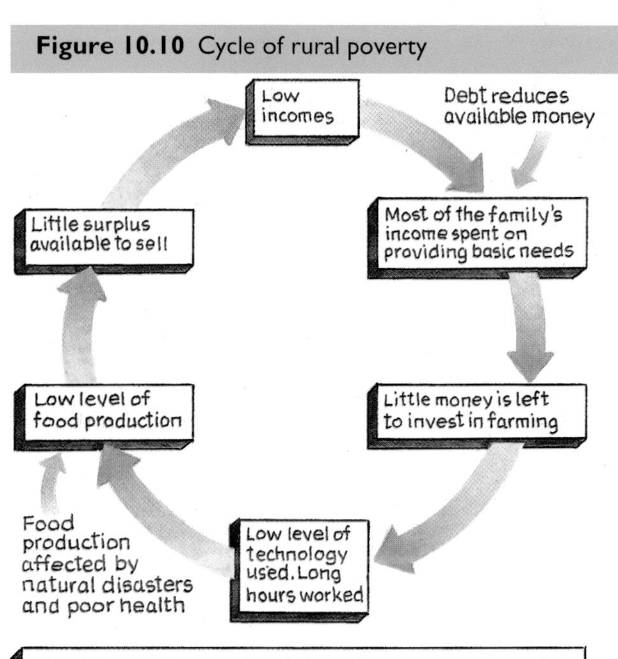

QUESTION BOX

1 Where are most of Brazil's industrial areas located?

2 Describe what is meant by a cycle of rural poverty.

3 Why is Brazil's south east described as the core region?

A Use Figure 10.8 to plot a series of maps to show the regional differences in Brazil. Use grade shading to draw your maps.

B Explain why the level of economic development decreases away from the core.

Economic progress

Brazil has the eighth largest economy in the world. It also has the world's largest foreign debt estimated at $123 billion. Since 1970 the Brazilian economy has become more industrialised. Brazil has exploited the large deposits of raw materials and the availability of a cheap workforce. Its government has encouraged transnational companies to invest in the Amazon by offering a range of incentives. Brazil has borrowed large sums of money to pay for several **prestige schemes** to be built, e.g. a new capital Brasilia, the Itpitu dam and the Trans-Amazonia Highway. These were intended to act as development poles for the region, attracting new industries and creating additional jobs.

In 1960, 80 per cent of Brazil's exports were **primary products**, e.g. coffee, rubber and mineral ores. Coffee made up half of the export total. By 1990, primary products accounted for only 10 per cent of the value of exports. The export of manufactured goods including high-tech equipment, aircraft trainers and cars had increased by 60 per cent by 1990.

Figure 10.11 Landsat photos of the rain forests surrounding the Rôndonia Development Project in western Brazil, during and after the building of the Trans-Amazonia Highway. a) Initial deforestation associated with the roadway, 1976. b) Extensive deforestation, 1981. Deforested areas appear blue/white, dense vegetation is red.

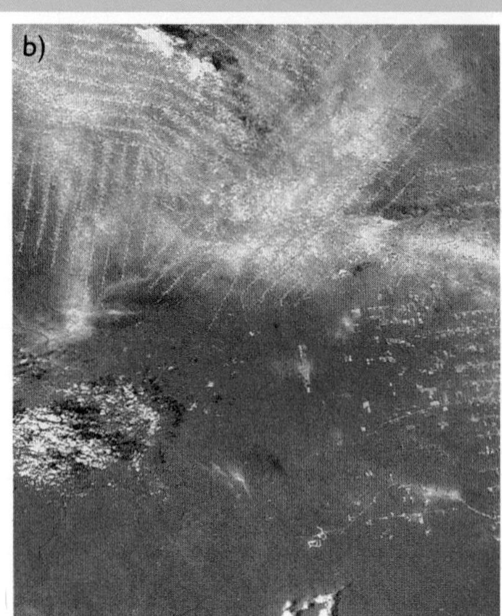

Amazon development

The Brazilian Government planned to develop Amazonia and has encouraged transnational companies to invest in the region. These companies pay little or no tax on their profits. Twenty development areas were established and connected by a new road network (Figure 10.12). Conservationists have been alarmed at the government's plans and lack of environmental protection in the region. The loss of the rainforest would affect the Amazon Indians' way of life and destroy a unique habitat for wildlife (Figure 10.13).

Figure 10.12 Amazon development plan

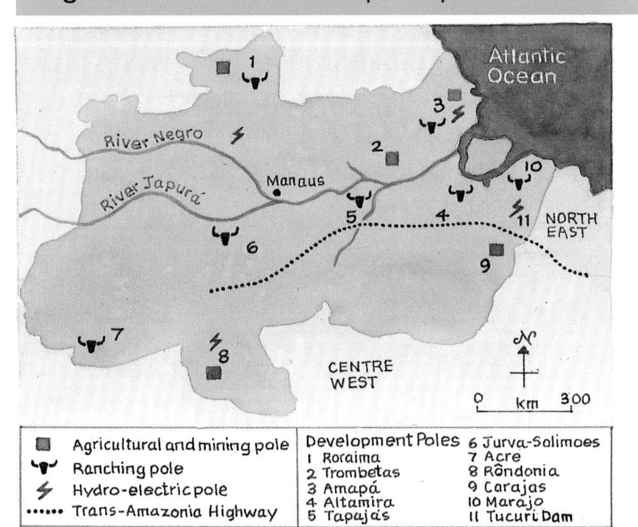

Others warn of global climate changes. The region has also attracted gold prospectors who use mercury and other chemicals to process the ore. These chemicals pollute the rivers. Uncontrolled forest clearance for cattle ranching has also damaged vast tracts of the rainforest.

Word box

prestige schemes A large project which cost vast sums of money, e.g. an HEP scheme or a manufacturing development which hopes to attract further development to the area

primary products Unrefined raw materials exported, e.g. iron ore, agricultural products

landsat images These are images from a remote sensing satellite orbiting at 900 km above the earth

Three development schemes in the Amazon

Trans-Amazonia Highway. This road was built in the 1970s to open up the Amazon region for development. It linked this isolated part of Brazil to the coast and to the neighbouring country of Peru. The highway was part of a network to attract people and industries from the crowded cities.

Iron mining at Carajas. Large deposits of mineral ores were found in 1967 in the Carajas region. The iron ore has a high iron content of 66 per cent and is easily mined using the open cast method with machines. A rail link to São Luis was built to export the ore to developed countries. Other deposits found include manganese, nickel, copper, gold, bauxite and tin. Most of the mineral ore will be exported. However, few local people are employed.

Resettling people from the shanty towns. With the building of the Trans-Amazonia Highway, the Brazilian Government offered shanty town residents money and a plot of farmland along the highway. When many arrived they found uncleared land, no services, no house and little help. Many returned to the city finding the work and trying to make a living impossible.

QUESTION BOX

1. List the type of economic developments planned for the Amazon.
2. What is meant by a Prestige Scheme?
3. Using the Landstat images, draw a sketch map to show the road pattern. Write a sentence to describe this pattern.

A Use the grid to estimate, as a percentage, how much of the forest has been cleared since 1986?

B Explain why conservationists are concerned about the large scale development of the Amazon.

Extension task: see worksheet 10A/B

Affordable development

Some developing countries have tried to become industrialised very quickly by investing large sums of money in a few prestige projects. These projects can make a country better-known, but they often do not benefit the majority of the population. They can make the development within a country uneven and encourage people to migrate from the poorer rural areas. The money borrowed often means that there is a large international debt to be repaid.

Intermediate technology

A cheaper alternative strategy uses local resources and traditional skills. This is called **intermediate** or **appropriate technology**. With help and advice, significant improvements can be achieved in peoples' standards of living (Figure 10.14).

Anokhi Textiles

This **co-operative** was a joint venture between traditional Indian craftsmen and western business. The textile company used local materials and maintained the village craft traditions in Jaipur, Rajasthan. Few machines are used and the work is labour-intensive. The textile designs are based on Indian patterns and sold in western countries. The co-operative helps to employ the traditional rural craftsmen and has begun to break the cycle of underdevelopment (Figure 10.15).

Johads

Johads are traditional water harvesting structures found in Rajasthan. They are made up of two 10 m embankments built using picks and shovels, along the contours about 1 km apart (Figure 10.16). They slow the surface runoff of water after the monsoon rains and form a small lake. The water is allowed to soak

Figure 10.13 a) Water tank, Brazil b) Paddy field farming, Kanchanaburi, India c) Syphoning water from an irrigation channel into rice paddy, Nigeria

Figure 10.14 Cycle of underdevelopment

Lack of investment. No new jobs created → Lack of jobs to keep in the area (→ migration out) → Local resources and skills underused → Decline in village industries → Village populations become more aged → Low incomes and standard of life → Lack of investment. No new jobs created

Word box

intermediate or **appropriate technology** A simple, low-cost improvement to a tool or way of making something which uses local materials and is done in a way ordinary people understand

co-operative A group of people with different skills working together

Johad Two embankments built on a hillside which trap monsoon rainwater into the ground and replenish wells. This prevents soil erosion and reduces water loss through evaporation. The water Johads are controlled by the village which built them.

Indian Government officials argued that the Johads were unsafe and took water away from the government-built dams. Officials believed that modern dams would solve the region's water shortage. These dams supplied electricity and irrigation to other areas where officials had responsibilities, and did not benefit the locals who lost their land and water supplies. Eventually, the government officials recognised the value of Johads and now help villages to build them.

Figure 10.15 Economic indicators for selected countries

Country	Annual income per person ($)	Adult literacy (% of population)	Cars in use	Energy coal equv/ person	Urban (%) population	% in farming
Bangladesh	200	29	15 000	0.09	14	59
France	24 990	99	22 million	5.15	74	6
India	340	50	15 million	0.3	27	62
Indonesia	560	74	74 000	0.47	33	56
Italy	19 020	97	24 million	3.95	67	9
Mali	270	27	19 000	0.02	27	85
UK	18 700	99	21 million	5.3	90	2

Figure 10.16 Indicators for selected countries

Country	Annual income per person ($)	Adult literacy (% of population)	Percentage employed in agriculture	Energy use per person (coal equv. in tonnes)
France	24 990	99	6	5.43
Italy	19 020	97	9	4.02
Mali	250	27	85	0.02
Bangladesh	240	36	59	0.08
Indonesia	980	83	56	0.38

QUESTION BOX

1 Look at Figure 10.14. State why each is an example of intermediate technology.

2 Why do many people leave or migrate, to escape the cycle of underdevelopment?

A Explain why intermediate technology can help rural areas in LEDCs improve levels of development.

B How could village cooperatives help break the cycle of underdevelopment?

Extension task: see worksheet 10C/D

SUMMARY QUESTIONS

1 The indicators in Figure 10.15 can be used to measure and compare levels of development

a) write in rank order each set of data

b) which countries are found at the top?

c) which countries are found at the bottom?

d) which countries are MEDCs and which are LEDCs?

2 Use Figure 10.15 to compare an LEDC with an MEDC using graphs to illustrate your work.

A Explain what is meant by a rich–poor divide at a global level and at a national level.

B Explain the difference between a Prestige Scheme and intermediate technology.

C Explain how a country could bring about improvements in the quality of life for its population.

D Is there a rich and poor divide in countries like Brazil or Italy?

unit 11

Energy and the Environment

Key ideas

- What is happening to the world demand for energy?
- How will the energy demands of the future be met?
- What impact does our use of energy have on the environment?
- What do we mean by sustainable development?

Supplying the world with energy

We all use energy: for cooking; for heating; when we switch on a light; or when we drive a car. In More Economically Developed Countries (MEDCs) energy is usually taken for granted. For example, few people think about which sources of energy are used to generate electricity when they switch on a television. However, this is not the case in many Less Economically Developed Countries (LEDCs) where gathering wood for the fire can take hours, or even days, of hard work (Figure 11.1).

Figure 11.1 Transporting firewood, Burkina Faso

Figure 11.2 Energy comparisons

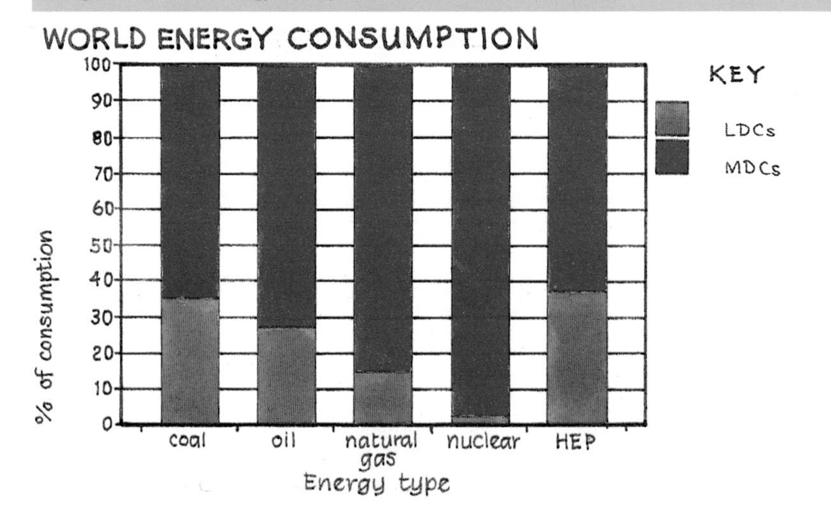

Figure 11.3 World energy consumption

Year	Consumption*
1965	4000
1970	5200
1975	6000
1980	6500
1985	7400
1990	7851
1995	8135

*all sources of energy worked out as if they were coal (million tonnes)

There are many different sources of energy. **Non-renewable** sources will eventually be used up whereas **renewable** sources will never be exhausted. Some types of energy are renewable as long as we manage them carefully; for example, we need never run out of wood as long as we plant new trees and wait for them to mature before they are cut down.

MEDCs use 75 per cent of the world's energy, although they account for only 26 per cent of the world's population. They do not produce all of this energy themselves. For example, in 1993 MEDCs imported over half of all the oil they used from LEDCs. Other energy comparisons between MEDCs and LEDCs are shown in Figure 11.2.

World energy consumption – the total amount of energy used – has risen considerably in recent years (Figure 11.3). This trend will continue as LEDCs develop their economies, as standards of living improve and as the world's population increases. This raises a number of issues which are considered in the following sections: when will our non-renewable sources of energy run out?; are there alternative sources of energy?; and what effect is our demand for energy having on the environment?

Word box

non-renewable A resource which will run out because there is a limited amount of it, e.g. coal, oil

renewable A resource which will never be exhausted, e.g. wind power, solar power

primary source Energy which has little, if any, processing before it is used, e.g. coal

secondary source Energy which has been made from a primary source, e.g. electricity, which can be made from coal or oil

QUESTION BOX

1 Make a list of the sources of energy used in your home for **a)** heating; **b)** cooking and **c)** lighting. For each source say whether it is non-renewable or renewable, and a **primary** or **secondary source**.

2 What does Figure 11.2 tell us about world energy consumption?

A Describe and explain the energy comparisons shown in Figure 11.2.

B Select and use an appropriate technique to present the statistics in Figure 11.3. Describe and explain what the statistics show.

Extension task: see worksheet 11A/B

Comparing two sources of energy: oil and wind power

Oil: a non-renewable resource

Oil forms from the remains of microscopic sea creatures. It is found in underground reservoirs, trapped by layers of **impermeable** rock. It is extracted by drilling a pipe down to the reservoir and it either comes up to the surface under its own pressure, or it is pumped out. It is then taken to a refinery where it is processed into many different forms e.g. petrol or diesel.

Oil has become increasingly important in the last 50 years and it now accounts for 40 per cent of the world's energy consumption. A number of reasons help to explain this: it has a relatively high **calorific value** (Figure 11.4); large reserves have been discovered; it is easy to transport, by tanker or pipeline; and it can be used to generate electricity. Oil is also the raw material for a wide range of manufactured products such as plastics, dyes and even animal foodstuffs.

However, oil has a number of disadvantages: for example, there is the risk of explosions and leaks when it is being extracted, transported and processed; and burning it releases a number of dangerous gases into the atmosphere (see pages 106–7).

At the present rate of consumption the world's reserves of oil will be used up in less than 50 years. However, it is extremely difficult to say exactly when oil will run out because new discoveries are being made all the time, and better methods of extraction are being developed. For example, BP and Shell are developing a new oil field, the Foinaven, 190 km west of Shetland. It contains four billion barrels of oil but this part of the Atlantic is 300 m deeper than the North Sea and the weather can be extremely rough, so a new way of extracting oil is being tested (Figure 11.5).

Figure 11.4 Calorific value of selected energy sources

Source	Energy content *
Peat	10–15
Lignite (brown coal)	15–25
Household coal	30–35
Oil	45
Natural gas	55

*Calorific value in megajoules per kilogram

Figure 11.5 Oil from the Atlantic

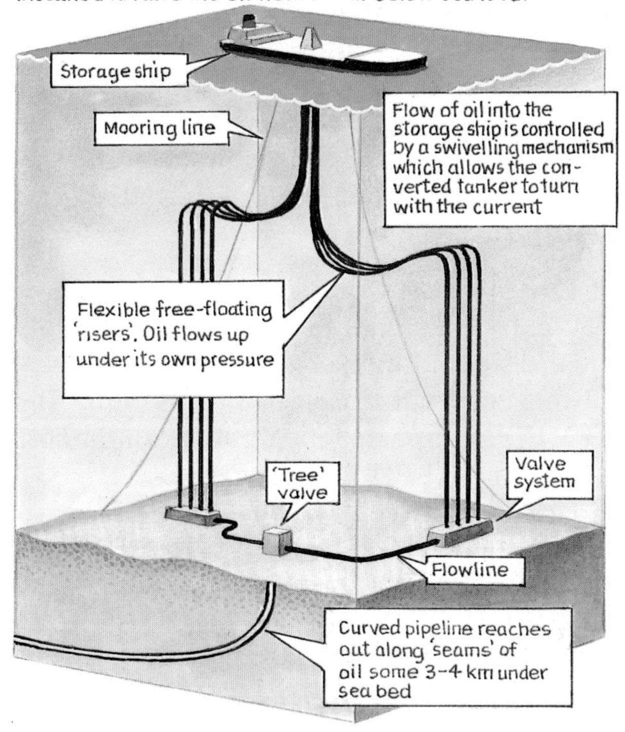

Wind power: a renewable source of energy

Wind power has been used for centuries: for milling corn; for pumping water; and for driving machinery. However, in recent years wind turbines have been developed to generate electricity. In the UK the government has agreed to subsidise 58 projects and there are already over 400 wind turbines in operation (Figure 11.6).

The main advantage of wind power is that it is a renewable source of energy and, as such, it could be used to replace non-renewable sources as they begin to run out. Also, it does not pollute the atmosphere.

However, wind turbines have their disadvantages. The wind does not always blow; they are noisy; they take up a lot of space; and they change the appearance of the landscape. It has been estimated that 7000 wind turbines are needed to replace one nuclear power station and that 200 000 would be needed to supply all of Britain's electricity.

Figure 11.6 Wind turbines in Carland Cross, Cornwall

Figure 11.7 Comparing oil and wind power

	Renewable or non-renewable?	Two advantages	Two disadvantages (one must be environmental)	Present importance	Future importance
Oil					
Wind power					

The British Government has committed itself to producing 3 per cent of the country's electricity from renewable sources by the year 2000. Two per cent already comes from HEP and there are other alternative sources of energy such as wave power and tidal power, so wind power (which currently produces 0.2 per cent of the country's electricity) is unlikely to become much more important in the short term.

Word box

impermeable A rock which fluids (e.g. water, oil) cannot pass through

calorific value The amount of energy in a given quantity of a resource

HEP (hydro-electric power). Electricity produced when running water, from a reservoir or river, is used to turn the blades of a turbine which is connected to a generator

QUESTION BOX

1 Copy and complete Figure 11.7, to compare oil with wind power.

2 Do you think wind farms, like the one in Figure 11.6, spoil the landscape? Give reasons for your point of view.

A What do you think are the main advantages and the main disadvantages of a) oil and b) wind power? Give reasons for the points you make.

B Do wind farms, in your opinion, have a positive or negative effect on the landscape. Give reasons for your answer.

Fossil fuels and the atmosphere

Fossil fuels – coal, oil and natural gas – account for 88 per cent of world energy consumption. However, they are also a major source of pollution because they release large amounts of harmful gases into the **atmosphere**.

Acid rain forms when sulphur dioxide and nitrogen oxides (given off by coal and oil, in particular) react with rainwater. It can kill trees, pollute rivers and speed up the **chemical weathering** of buildings. Winds can blow acid rain for hundreds of kilometres, so not all of it falls on the country which produced it; for example, 20 per cent of the acid rain produced by the UK falls on Norway and Sweden (Figure 11.8).

Figure 11.8 Acid rain in Europe

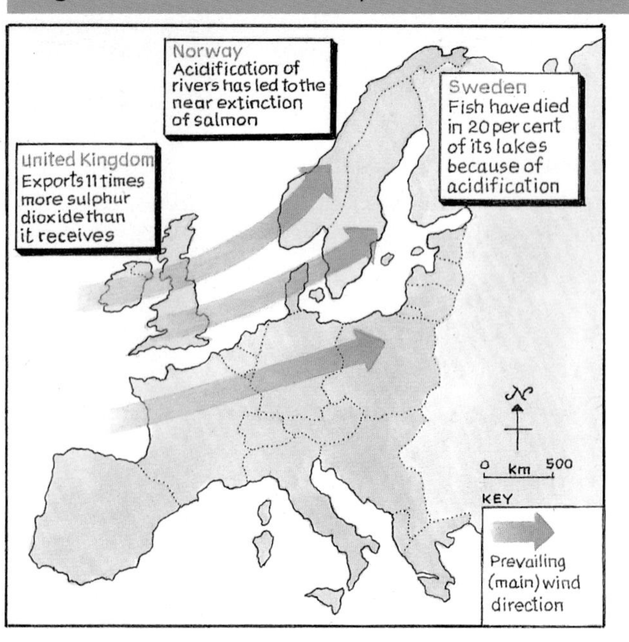

Burning fossil fuels also releases carbon dioxide which is a 'greenhouse gas' – this means that it stops heat escaping from the atmosphere (Figure 11.9). By adding carbon dioxide to the atmosphere we are trapping more heat and this could lead to **global warming**. Higher temperatures could help some parts of the world but not others: for example, cool regions would have a longer growing season but if the polar ice caps melted sea levels would rise and some places would be flooded. There is already some evidence that global temperatures are rising but, as yet, no one is really sure.

Figure 11.9 The greenhouse effect and global warming

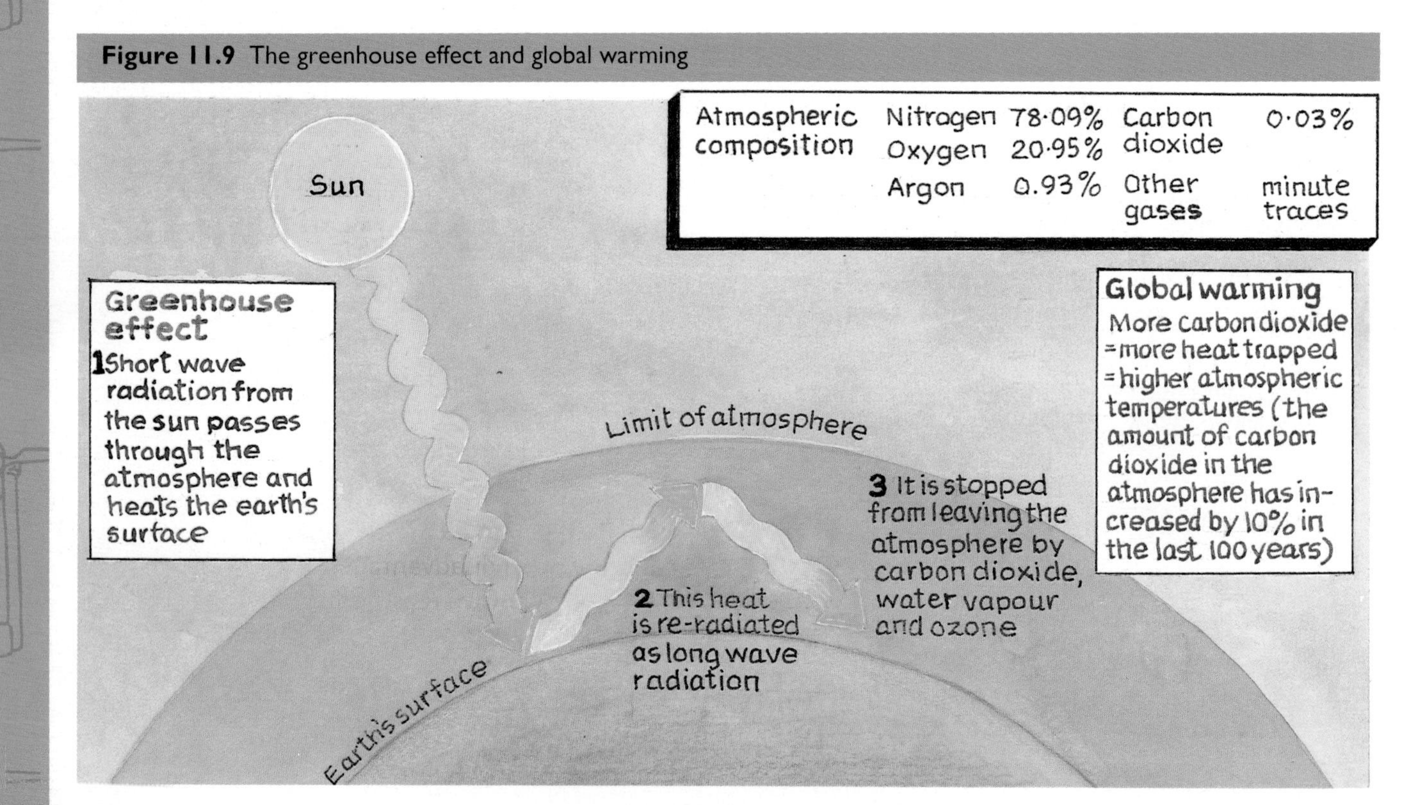

Figure 11.10 Photochemical smog in southern England, July 1994

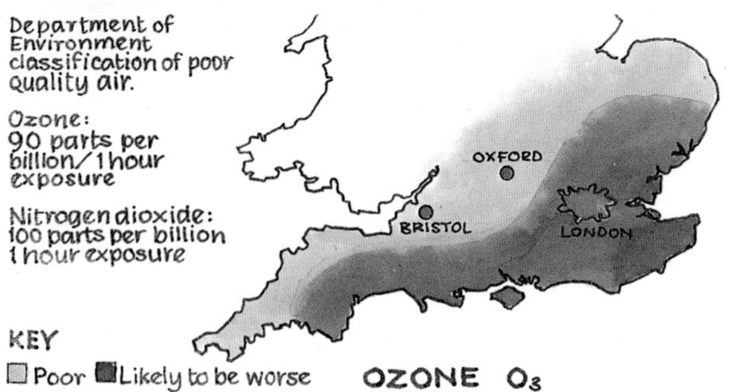

Another problem is **photochemical smog** which can often be seen as a brown haze and is caused mainly be vehicle exhaust fumes. It is most likely to form when the weather is calm (because there is little wind to blow it away) and warm (because the higher the temperature, the more rapid the chemical reactions). This explains why, for example, it has become a health hazard in parts of the UK in the summer months (Figure 11.10).

QUESTION BOX

1 What is acid rain and what effect has it had on Norway and Sweden (see Figure 11.8)?

2 Draw and label a diagram to show how acid rain forms.

3 Look at Figure 11.9. What is the greenhouse effect? What is global warming? What might be causing global warming?

A Why are international agreements necessary if we are to tackle problems such as acid rain and global warming?

B Explain why global warming could be an advantage for some parts of the world but not for others.

Extension task: see worksheet 11C/D

Some things can be done to tackle these problems, for example:

- acid rain can be controlled by fitting filters to power station chimneys (Figure 11.11), or by adding lime to polluted water;
- the amount of carbon dioxide released into the atmosphere could be reduced if we saved energy;
- smog would be less of a problem if people shared cars or used public transport.

However, many of these solutions are expensive (electricity prices could rise by 20 per cent if filters were fitted to all power station chimneys), or unrealistic (many people have to use their cars because the public transport system is not good enough). Also, anti-pollution measures could mean the closure of power stations, mines and factories which would cause unemployment.

Figure 11.11 Meremere coal fired power station

Word box

atmosphere The layer of gases which surrounds the earth

acid rain Rainwater which has become acidic because of pollution by chemicals in the atmosphere

chemical weathering The breakdown of rocks and building materials because of chemical processes

global warming An increase in the temperature of the earth's atmosphere

photochemical smog A fog or haze caused by the reaction of chemicals, water vapour and sunlight

Sustainable development

The earth's energy **resources** are being used up at an ever faster rate because of population increase and economic development. Non-renewable resources, such as oil, could run out before we have developed replacements. Also, our increased use of resources is causing pollution – locally, nationally and globally.

People have different ideas about what would be the best sources of energy for the future. However, most people agree that we should aim for sustainable development. Sustainable development has three main parts to it: people, the economy and the environment (Figure 11.13). These are linked together, e.g. if the population increases there is a greater demand for jobs which places a greater pressure on the environment.

Sustainable development means the sort of development the earth can cope with. It accepts that some development is inevitable because, even if the world's population stabilises, the gap between the world's rich and poor coutries is so great (see pages 92–3) that LEDCs will want to catch up. However, sustainable development does not allow anything which would cause lasting damage to the earth and its ecosystems.

Figure 11.12 Department of the Environment energy saving poster

Word box

resources Things which are of use to people
sustainable development Development the earth can cope with

QUESTION BOX

1 Suggest ways in which your school could reduce the amount of energy it uses.

2 Why might a LEDC want to use a newly discovered coalfield even though this will cause a lot of pollution?

A Describe and explain three things you could do to save energy. How would your suggestions affect your lifestyle?

B Find out about the Earth Summit in Rio de Janeiro in 1992. An Internet search will yield a great deal of information but remember to evaluate a site before using it, e.g. who wrote it? Have they got a particular point of view? Write a short account which describes: who took part; what was agreed; what some of the countries disagreed about; and what progress has been made.

Sustainable development needs everyone to be involved. For example, there is little point in just one house in a street using energy saving light bulbs, or one country reducing the amount of fossil fuel it uses. Here are some examples of what is, or could be, happening:

- individually, switching off unnecessary lights not only saves money but also conserves fossil fuel and cuts atmospheric pollution;
- nationally, governments could pay for research into alternative sources of energy. For example, in order to reduce its oil imports, the Brazilian Government in 1975 set up 'Proalcool', a plan to use alcohol made from sugar-cane instead of petrol. Now, a quarter of all Brazilian cars are designed to use this fuel;
- internationally, governments could reach agreement about issues such as atmospheric pollution. For example, in June 1992 the first Earth Summit was held in Rio de Janeiro in Brazil. Very little was decided but it did bring together 130 of the world's leaders and they did agree to hold further discussions about sustainable development.

Figure 11.13 Earth from space

SUMMARY QUESTIONS

1 Carry out a survey to compare what ten of your friends, and ten of your friends' parents, think about environmental issues. Write five or six short questions e.g. Are you worried about global warming?, or Do you think wind power is a good idea? Present your results in an appropriate way – perhaps as a series of graphs. Describe and explain any differences between the two age groups. Could your results be misleading in any way?

A Find out about Local Agenda 21 in your area. Your local library and/or Council should be able to help you with information. (County, City and District Council web sites are a good place to start a search.)

a) What is Local Agenda 21?

b) What has been achieved so far in your area?

c) What is planned for your area?

d) Are there any ways in which you could get involved?

INDEX

KS3 PROGRAMME OF STUDY: ANALYSIS

The following analysis demonstrates how Core Geography meets the requirements of Curriculum 2000 Geography. Page numbers are given in *italics*. Ideas as to how the different units and sections of the text can be put together to support KS3 plans varying in theme/issue/place approaches are given in the Teacher's Resource Pack. These are examples only. Many more opportunities exist in the book.

GEOGRAPHICAL ENQUIRY AND SKILLS

Geographical enquiry and skills are an integral part of the text. As such, the following list indicates only a few of the opportunities for their development. Other opportunities, particularly for geographical enquiry and OS mapwork, are presented in the Teacher's Resource Pack.

1a–f *43, 109*
2a *See Word Boxes*
2c *91*
2d *13, 33, 43, 71, 91*
2e *25, 33, 63, 83, 103*
2f *43, 71*
2g *51*

KNOWLEDGE AND UNDERSTANDING OF PLACES

3a–e, 6a
Unit 7 *Japan*, **Unit 8** *India*
See also the references to Italy and Brazil in the index.

KNOWLEDGE AND UNDERSTANDING OF PATTERNS AND PROCESSES (4a–b)

6b *4–13*
6c *14–23*
6d *24–33*
6e *34–43*
6f *44–53*
6g *54–61*
6h *82–91*
6i *92–101*

KNOWLEDGE AND UNDERSTANDING OF ENVIRONMENTAL CHANGE AND SUSTAINABLE DEVELOPMENT 5a–b:

6j 22–23, 40–41
6k Unit 11 102–109

Note also
7a local, LA earthquake *8–9*
regional, Hokkaido vs Kanto *66–7*
national, Italian car industry *86–87*
international, world trade *90–91*
global, greenhouse effect *106–07*
7b UK *26–29, 83*; EU *30, 91*
7d *42–43, 106–107*